Blurring the Edges
by
Steve Dunthorne

ISBN: 978-1-914933-30-1

Copyright 2022

All rights reserved. No part of this publication may be reproduced, stored in a retrieval system, or transmitted in any form or by any means, electronic, mechanical, photocopy, recording or otherwise, without prior written consent of the copyright owner. Nor can it be circulated in any form of binding or cover other than that in which it is published and without similar condition including this condition being imposed on a subsequent purchaser. The right of Steve Dunthorne to be identified as the author of this work has been asserted in accordance with the Copyright Designs and Patents Act 1988. A copy of this book is deposited with the British Library.

All advice and expertise offered in this book is given with best intention and, to the best of the author's knowledge, is correct and safe. As such, any actions undertaken by the reader because of information gleaned from the book are at the reader's own risk and must be in accordance with all current regulations pertaining in the country of residence. If there exists any doubt or lack of sufficient knowledge or training of the reader to perform said actions, professional advice should first be sought before embarking on any such procedure.

Published By: -

**i2i**
PUBLISHING

i2i Publishing. Manchester.
www.i2ipublishing.co.uk

Buying, assembling, and teaching myself to use a Tormach® 770MX CNC milling machine: my journey from distinctly novice to relative competence

# Acknowledgements

I have been made to feel profoundly welcome in the publishing world by Lionel Ross, who heads up i2i Publishing in Manchester, UK. Lionel was quick to forgive my shortcomings as a complete beginner in the art of writing. Moreover, it was he who first delved into my original draft and gave me the thumbs up, so I shall always remain grateful for his confidence in my idea.

A significant thank you is owed to Roy Clayton, my assigned editor, who contended with my previously undiagnosed aversion to hyphenation and bravely ploughed through my script to reveal numerous errors and teaching me some subtle nuances of our beautiful language along the way. I fully appreciate it may not have been his subject matter of choice; nevertheless, his input was vital and I enjoyed repairing the book with his excellent suggestions.

I would like to add my thanks to the team at Tormach in Wisconsin USA, for their continued and unwavering assistance and advice on operating my machine, and specifically to Andrew Grevstad, their Business Development Director, for giving me permission to use some of their in-house brand imagery which I trust you will agree has added to the book's authenticity and vibrancy.

Finally, I would like to offer my sincere thanks to my lovely wife Sarah, who continually brought me cups of hot fresh tea accompanied by a genuine smile, when I was spending long hours in the garage getting to know my machine, and later when I was at my desk tapping away at my keyboard, endeavouring to encapsulate my story in a way which I dared to hope would be interesting.

# Foreword

From the very beginning, our mission at Tormach has been to 'enable your ideas'. We've chosen to do that by focusing on creating affordable high-value CNC products so that digital manufacturing that was once only available to large factory concerns can be within reach for small manufacturers, educators, hobbyists, and tinkerers.

I've enjoyed meeting and getting to know Steve Dunthorne as he has embarked on his CNC journey. It's a rewarding part of my job to interact with passionate life-long learners, and Steve embodies this like very few others. The narrative he has shared in this book is an invaluable perspective for anybody looking to dive into learning CNC, and not without a few sage tidbits that were new to an experienced insider like myself.

These days, in the modern information economy, there is so much sharing – via YouTube, social media, etc. It can be hard to separate the noise from valuable perspectives. Steve's user journey, warts and all, is authentic and detailed and thoughtfully considered – a true gem for anybody considering taking the next steps towards learning CNC.


Andrew Grevstad
Business Development Director
Tormach Inc., WI, USA

March 2022

# Preface

I never had plans to write a book, least of all a book about a milling machine. It happened quite by accident. I purchased a reasonably large machine, by hobbyist standards, and the engineering project which became of it surprised and amazed me in equal measure. A new subject for me, the learning curve was relatively steep although immensely satisfying and, on realising I had achieved something useful, I suddenly felt driven to share the experience. I have tried to encapsulate everything I did along the way. If you are an engineering enthusiast, a hobbyist, or maybe even a 'start-up' entrepreneur with an exciting idea toying with the notion you might like to buy a CNC milling machine, to do some prototyping, or maybe even the beginnings of 'small town' production, then this book should prove a useful read. I certainly hope so.

Through a chance meeting at work with a very clever man called Rob Joseph, we got talking about desktop 3D printers; how technically developed they had become, how much more affordable they were in recent years, and how much I jolly well ought to buy one! For my 'day job' I don't work in any engineering capacity, but my background certainly had me starting out that way, an electrical engineering degree and a sound understanding of basic engineering principles and workshop techniques; over the years my garage had become the cliché cave which now sported a small manual lathe, an even smaller manual mill and recently I had dabbled with TIG welding – after all, you can teach yourself anything these days, just look it up on YouTube, right?

Rob had caught me unawares at just the right moment and, succumbing to his friendly cajoling, I placed an order for a brand-new Prusa Mk3S desktop 3D printer and began assembling it as soon as it arrived – a wonderful machine which continues to surprise in its usefulness to this day. That's another story, not the reason for this book I hasten to add, but the point being a 3D printer opens a whole new world of 3D modelling. Until this point in time, if I wanted to fabricate a 'thing', I would draw a quick pencil sketch, power up the mill, and start cutting and drilling.

This had served me well for decades and now suddenly I was entering a very different era. Long ago, back in the late 1970s in fact, I had been taught how to pen a proper technical drawing. It was called a 'third angle projection'. I had no experience of what computers now offered in the modern-day technical drawing scene; I had never worked in that environment. Rob had introduced me to some software, free to download for the hobbyist or personal user, called Fusion 360, from a company called Autodesk. This was literally eye-opening and, with plentiful use of YouTube videos and Google searches, I enthusiastically taught myself the basics of how to make a drawing of the next 'thing' I was going to make. Only this time, I would be modelling it in three dimensions, not three projections.

Fifteen months after our first meeting I bumped into Rob again, enthused about the printer, and began asking him lots of questions about Fusion 360 as I knew he was quite the expert on this. He could tell I was hooked and that's when he suggested,

"Why don't you get yourself a CNC milling machine?" And this book tells the story that followed.

It is disconcertingly easy to put a great deal of time into research and spend literally days looking at all the purchase options, what the review sites are saying, what comments are prominent on enthusiasts' online chat forums and what people are doing on YouTube. I had some machine manufacturers' brand names in my head following my chat with Rob. I still didn't know which I would eventually buy, but I knew I had a few more questions for him and I also needed to get a book to find out more. This was not going to be a simple purchase.

What struck me most, when I began looking for more information, was a lack of resource originating from Europe in general, and the UK specifically. Hobbyist CNC milling, on the internet at least, is overwhelmingly dominated by our colleagues in the USA and, consequently, the imperial measuring system seems to prevail. I come from an era when both systems were taught with equal validity, and I can speak in 'inches and thou' just as I can in 'millimetres and microns.' Nevertheless, the expectation was I would be operating my milling machine in the metric system. That's an obvious choice if you live in the UK, and my natural way of thinking. It is more testing for us Europeans to learn from American sources because of this difference so if I did write this book, I reasoned, I would aim it squarely at engineering enthusiasts who would be installing their machines to operate in metric, if only because that is exactly what I had just accomplished.

Looking to purchase books about CNC milling and at the range of material on offer, there appear to be two distinct target audiences. On the one hand, there are books which start off explaining the Cartesian coordinate system, how we label the three dimensions the X,

Y and Z axes, how (in the last century) they began controlling milling machines numerically using punched ticker tape much like a Telex machine from the 1970s and how this had morphed into computer numerical control with the advent of mainframe computers. This is a basic history; laudable, but not what I needed.

On the other hand, there are books out there which are brilliant, technically speaking, going into detail which I could never dream of writing. I bought these books, but I've never properly read them. They can often be too technical for me, so I can only read so far before I glaze over, or they are intended as references to which I might wish to return when I need to look up a particular issue, rather than just read from cover to cover for entertainment. There are some books which go into bewildering detail about G-code and attempt to teach you how to program a CNC controller from scratch, a skill which has very much been superseded, dare I say usurped, by advances in software, so is arguably of declining importance.

I wanted a book which was more practical, which would help me start this project at a level which was not trying to cover all bases, and at the same time was accepting of my having an appreciable level of old-fashioned engineering knowledge, whilst guiding me to the new. The question arose, was I asking too much?

I was going to need to know more about G-code, the program code which controls the motions of the cutting machine. In comparison with CNC, learning desktop 3D printing is relatively straightforward. Whether you download items you want to print from well-established internet sites such as 'Thingiverse.com' or make your own designs using modelling software such as Fusion 360®, you end up with a file (maybe an STL or latterly a 3mf file type) which you upload to special software, something called a slicer, and then this generates the G-code specific to your printer to make the part. You simply instruct the slicer how thick you want each printed layer to be, and what sort of plastic you are using (and hence what temperature to set the extruder) but, other than that, you don't really have to know anything at all about how it does it. You certainly don't get to learn anything about G-code, even though your printer is controlled by it. That's all taken care of by the slicing software.

As I was about to find out, it is not so straightforward with CNC machining and you certainly do need to know a few things about G-code. You won't need to learn how to write the code to cut some material into a complex curved shape, to drill a hole, to chamfer an edge, or to make a thread. In all these examples, and more, the latest computer-aided manufacturing (CAM) software writes all the G-code for us. But there are elements of G-code we should master, and so we must advance from 'designing' in three dimensions to learning how to 'make' in three dimensions and, to help us, there is extra software that is normally a sub-section of the 'design' package. In my case, I had become quite attached to Fusion 360 and it would become a natural progression for me to learn how to use the 'manufacturing' space within it. Later, I write about getting to grips with CAM rather than learning the intricacies of G-code. Learning to use your chosen CAM software is akin to programming in a 'high-level' language. Those engineers writing G-code in the 1960s were operating at the 'assembly language' level. We don't need to do that any longer. As for teaching yourself how to program CAM in Fusion 360, the internet is clearly the most suitable resource. That, and just experimenting with it.

The books I found, on computer-aided manufacturing, tended to cover the subject from the basics and I don't think you need to do this if you are going straight in at 'high-level'. Also, G-code is not only concerned with instructions to cut parts; it additionally covers the setup of the machine before it can even begin to think about doing any cutting. Once again, there are aspects of G-code which are vital to know, and some less so. If I was to write a book, I determined, I would attempt to strike a balance and cover those parts I needed when I learned how to use my machine, but no more. I certainly would never try to write a book on CAM – I would be unqualified to do so.

By now you'll appreciate I was having trouble finding a book to help me start my project. From my perspective, there was a gap in the market. I took on the challenge and assembled my new machine using all the resources I could muster and, as I progressed and learned about the subject of CNC, my own machine grew in capability on account of my growing knowledge. The moment was reached when it dawned on me my own experiences might be the building block of the book I originally sought.

This book assumes you have some general engineering knowledge. You may have your own desktop 3D printer too. You probably have some other workshop tools which are manually operated. Most of all, you now want to take the steps of choosing your own CNC milling machine, getting it assembled, and successfully up and running, with the minimum of fuss. This book is a genuine and authentic technical journey just like the one I took, in my personal transition from CNC newbie to 'yes I can actually do this.'

That journey is certainly not always easy. There are engineering difficulties to overcome, snags along the way; all solvable but they *will* test you. This is the story of my journey from researching and placing my order right through the 'build' process all the way to 'making chips', as the Americans say. If this book assists you in taking a similar path, then it has done its job.

When I assembled *my* machine, I was unable to find a book like this.

# Introduction

This is a chronological record of my journey starting out as a CNC milling 'complete novice' to someone with a reasonable appreciation of the subject. The journey was genuine. By that I mean I didn't merely teach myself about the subject just so I could pass on what I found out; I purchased a real machine, assembled it in my garage, and taught myself how to use it. I can make parts now which I could only have dreamt of making before.

I begin with deciding which machine I should buy. Important deal-breakers such as size, weight, and power supply requirements come into play. Once that decision is made, it is not as simple as you might imagine going ahead and placing an order. There are so many optional extras and add-ons available to you once you have a machine in mind, and it is well worth taking the time to consider each one with care. I made my personal decision to get a Tormach® 770MX CNC mill from the USA and I take you through the extensive process of finalising the exact specification in the second chapter.

Another section is devoted to the inordinate amount of time spent making preparation for the imminent delivery. There is a vast amount to do in readiness for the arrival of the machine: providing mains electrical supply, sourcing special tools, making space, preparing for seriously large and heavy crates arriving, the list goes on, and you can learn from my mistakes.

Throughout the book you will see me describe my 'top tips' and these are simply those aspects where I learned the hard way, but you don't need to. I realise these are often specific to the Tormach® machine I was assembling, as that's what this project became about, but much of the advice is generic and will be relevant for other types.

Learn about putting all the pieces together to produce a fine and hopefully long-lasting piece of electromechanical wizardry, the likes of which would not have been possible in the domestic environment not twenty years ago. It's a satisfying personal accomplishment.

I make the point: this is not a book about 'how to assemble a Tormach® CNC mill' – the Tormach® supplied Operator's Manual does that perfectly well – instead this is a book looking to guide you during the entire process, keeping you aware of some of the pitfalls and 'gotchas' along the way. I hope, and believe, there is synergistic value in reading this book side by side with the manufacturer's own instructions.

I have written in chronological order as, part way through the assembly process, there are many times when you need to power up the machine to get things into the correct position or test how things are progressing, so you will be gradually building up knowledge of the control system as you go. Only when you start taking some first tentative cuts with the fully completed machine, it's possible you may suddenly realise you know hardly anything about what to do next. It's exciting stuff but you can't afford to cut corners and damage the machine so I try to describe some early gentle steps to ease you into the basics in a comfortable manner, and hopefully some of the guidance will keep you safe and the machine unscathed.

Unlike using a desktop 3D printer, you cannot hope to operate a CNC mill without gaining some basic knowledge of the code which controls everything, the G-code. I attempt to find the middle ground in explaining just enough of the G-code you will come across, and not over-complicating things with code you might never see. It is true to say the modern CAM (computer-aided manufacture) software systems have largely taken over the 'clever bit' of someone genuinely writing code to cut complex shapes, so a gentle introduction is all that is needed. It is important to realise some G-code is instructional, telling the machine to 'do' things, but a lot of it is setting up modes for the machine to be 'in'.

And then we come to the wide-reaching and absorbing subject of tooling. This is where you get to choose what you want to make with your machine, so it makes sense you get to choose which tools you will likely need; the various types of tool holders, collets, cutting tools, and engraving tools. The Tormach® is excellent at engraving, especially in conjunction with the CAM package within Fusion 360.

I acknowledge some readers will naturally shy away from the mathematics of CNC milling but I implore them to stick with me on this important aspect. After all, the subject of feeds and speeds is key to truly understanding how well your mill is working. Those readers more comfortable with mathematics will recognise I have carved a gentle path to explaining these fundamental parameters and how we can use tool manufacturers' published data to successfully set up our CAM programs.

A chapter is given over to considering many of the issues surrounding work holding and the techniques you can employ. I think this is a difficult subject when it comes to complicated parts because there is often not one single solution. Choosing which way to go is where the human skill is still required, which pleases me a great deal. I start with some basic and simple techniques and then look at a slightly more challenging example which I had to solve, illustrating my thought processes.

I will write briefly about 4th axis machining. Tormach®'s microArc4 option is an excellent device but needs appreciable care and attention installing it the first time, and this is discussed. As for making parts using the 4th axis, I am still very much a beginner so I don't have much advice to offer yet. The journey never ends but the challenge is inspiring, all-consuming and one I can't believe I had the good fortune to embark on.

I finish with a list of all the clever people out there who helped me learn how to do this. Such is the beauty of the knowledge, openness, and generosity residing within the World Wide Web and I thank them all profoundly for unwittingly helping me write this book.

# Safety

This is a book about installing a CNC milling machine into a domestic environment, not an industrial one, and such a scenario could cause an increased danger to life. Health and safety regulatory standards are well documented for the workplace, with the full force of the law behind them, but this is certainly not the case in your private garage.

In truth, we can pretty much do as we like behind our own closed doors but that doesn't relinquish any of us from the 'duty of care', incumbent upon us all, we must show to others.

Industrial machinery is profoundly dangerous in the wrong hands, so it is necessary for anyone venturing into this arena for the first time to educate themselves using all resources available. It is also vital to appreciate in the domestic situation, untrained, possibly young, and inexperienced, persons might easily gain access to dangerous parts of the setup, something most unlikely in an industrial setting.

Any work on the 'mains' electrical supply demands some knowledge of the subject. If you do not have this, then you must sanction the help of a qualified electrician. If you do have the knowledge, then you must still get a qualified person to 'sign off' the work completed, particularly regarding 'earth fault loop impedance' testing. Any advice and technical information given in this book is generic and must be read in conjunction with the current regulations in your country of residence.

A pneumatic air supply presents obvious dangers which, unfortunately, are not in the least bit obvious to the very young, so some effort to make this aspect as safe as possible is worth additional expense and I offer a particular suggestion later in the book.

During the lifting phase when assembling the heavy items, specific care must be taken to ensure safe operation of the hoist. The weights involved mean any accident could result in a danger of serious injury or even death.

In the non-industrial setting, whether to choose the option of an electric door lock on a CNC mill is down to personal choice. If no lock is fitted, then it is important to be extremely careful not to inadvertently place fingers near the dangerous bits when the machine is operating. If using pumped coolant, the doors would naturally be closed to stop any splashing onto the floor, but the temptation to leave them open when dry cutting is strong, so industrial eye protection is an absolute necessity. These are recommended even when shielded behind the Perspex windows in the sliding doors. Carbide cutting tools breaking at high speed will have extraordinary kinetic energy.

Lubricating oils and coolant chemicals stored near the machine must be kept safe from children. Once the mill is working, a means to collect and dispose of the swarf produced will also be needed. Use of an electric bandsaw to cut up material stock prior to machining also poses the opportunity for dangerous situations to occur, so 'shop cleanliness' is a desirable habit.

Finally, any self-respecting milling machine owner-operator will build up a suitable collection of cutting tools fitted in tool holders. Any rack or tray storage system always presents the 'sharp ends up' so care is needed to ensure these are beyond the reach of unplanned visitors, and suitably positioned away from the operator's own carelessness. Some of the tools are incredibly sharp.

Please be careful. Proactively looking out for danger is something which should NEVER stop.

# Contents

## Chapter 1  Choosing the Tormach® 770MX ..................................... 27

1.1 The type of CNC controller .................................................................. 27

1.2 Choosing the Tormach® 770MX .......................................................... 31

    1.2.1 Price ................................................................................................ 31

    1.2.2 Size of the overall milling machine ........................................... 32

    1.2.3 Size of the parts it can make ....................................................... 32

    1.2.4 Electrical power supply requirement ........................................ 33

    1.2.5 Stepper motors or Servo motors ................................................ 33

    1.2.6 Tool holder type ........................................................................... 34

1.3 Purchase options .................................................................................... 36

    1.3.1 Automatic Tool Changer (ATC) ................................................. 36

    1.3.2 MicroARC4 – work holding in the 4th Axis ............................. 38

    1.3.3 PathPilot® Console ...................................................................... 39

    1.3.4 Flood Coolant ............................................................................... 42

    1.3.5 Tramp Oil Skimmer .................................................................... 43

    1.3.6 Automatic 'Ways' Oiler .............................................................. 44

    1.3.7 Electronic Tool Setter (ETS) ....................................................... 45

    1.3.8 Work Offset Probe ....................................................................... 48

    1.3.9 Work holding ............................................................................... 50

    1.3.10 Tool holding ............................................................................... 51

    1.3.11 USB M-Code I/O Interface Kit ................................................. 52

    1.3.12 Lifting bar kit .............................................................................. 52

## Chapter 2  Machine specification ........................................................ 53

2.1 Tormach® 770MX specifications in detail ......................................... 54

2.2 Feeds and speeds – first thoughts ....................................................... 55

2.3 Spindle belt drive settings .................................................................... 56

2.4 Table size and machining volume in x, y and z ................................ 56

2.5 Synchro tapping and thread milling .................................................. 56

2.6 Coolant system ....................................................................................... 57

2.7 PathPilot® console controller .............................................................. 57

## Chapter 3   Preparations for delivery .................................................. 59

   3.1   Allow enough space ........................................................................ 59

   3.2   Download the Operator's manual (pdf) ........................................ 60

   3.3   Subscribe to Tormach®'s YouTube channel ................................. 61

   3.4   Electrical supply ............................................................................... 61

      3.4.1   Dedicated supply? .................................................................. 62

      3.4.2   Step-down isolating transformer ........................................... 66

      3.4.3   Connecting the transformer primary to the mains ............. 71

      3.4.4   Connecting the transformer secondary to the Tormach® .. 73

      3.4.5   Distribution board Mains Circuit Breaker: type B or type C? .. 76

      3.4.6   Power supply for a Tormach® 1100MX on 230 volts ........... 78

   3.5   Pneumatic air supply ...................................................................... 80

   3.6   Spirit Level ....................................................................................... 83

   3.7   Concrete floor load calculation ..................................................... 86

   3.8   Coolant .............................................................................................. 87

   3.9   Coolant Refractometer .................................................................... 90

   3.10   Lubricating oil for the machine slideways ................................. 91

   3.11   Pneumatics lubricant ..................................................................... 92

   3.12   Engine hoist .................................................................................... 92

   3.13   Pallet truck ...................................................................................... 95

   3.14   Reciprocating power saw ............................................................. 96

   3.15   Allen keys in your toolkit ............................................................. 97

   3.16   Metal snips, sometimes called tin snips ..................................... 98

   3.17   American screwdriver ................................................................... 98

## Chapter 4   Delivery Day ................................................................... 99

   4.1   Out of balance crates on pallets ..................................................... 99

   4.2   Position crates carefully in the garage ready for dismantling .. 100

   4.3   Plan for rubbish disposal ............................................................. 101

## Chapter 5   Assembly ........................................................................ 103

   5.1   Start of the build ............................................................................ 103

   5.2   The stand ......................................................................................... 104

- 5.2.1 Front access .................................................................................................. 104
- 5.2.2 Left side access ............................................................................................ 106
- 5.2.3 Right side access .......................................................................................... 106
- 5.2.4 Behind the machine ..................................................................................... 106
- 5.2.5 Clearance above ........................................................................................... 107
- 5.2.6 Fixing the feet to the machine stand ........................................................... 107
- 5.3 The milling machine itself ................................................................................. 108
  - 5.3.1 Releasing the mill from the pallet ............................................................... 110
  - 5.3.2 Lowering the mill onto the stand ................................................................ 111
- 5.4 PathPilot® controller and first 'power up' ........................................................ 112
- 5.5 Automatic oiler .................................................................................................. 113
- 5.6 Pneumatics ......................................................................................................... 114
- 5.7 Flood coolant system ......................................................................................... 117
- 5.8 Automatic tool changer (ATC) ......................................................................... 120
  - 5.8.1 They changed the design of the mounting bracket ..................................... 123
  - 5.8.2 Electrical interference on the ATC USB cable ............................................ 124
  - 5.8.3 Air blast nozzle problem ............................................................................. 125
- 5.9 Building the enclosure ....................................................................................... 128
- 5.10 PathPilot® console assembly ............................................................................ 133

## Chapter 6    PathPilot® ............................................................................. 135

- 6.1 General synopsis ................................................................................................ 136
- 6.2 Status line ........................................................................................................... 137
- 6.3 Zeroing the axes manually ................................................................................ 137
- 6.4 Current tool indicator box ................................................................................. 137
- 6.5 Go to G30 button ............................................................................................... 137
- 6.6 Status tab ............................................................................................................ 138
- 6.7 Store Current Tool command button ............................................................... 138
- 6.8 Avoiding a crash ................................................................................................ 138

## Chapter 7    First cuts ................................................................................. 141

- 7.1 So where to start? ............................................................................................... 141
- 7.2 Work offsets ....................................................................................................... 142
  - 7.2.1 G54 work offset ............................................................................................ 146

- 7.2.2 Flip for the second setup ............................................................................... 148
- 7.3 Tool offsets ........................................................................................................ 150
  - 7.3.1 G43 code ................................................................................................... 152
  - 7.3.2 Tool libraries (real and not-so-real) ........................................................ 152
- 7.4 G30 code ............................................................................................................ 153
- 7.5 Conversational programming ........................................................................ 155
- 7.6 Post processor .................................................................................................. 159
- 7.7 Fusion 360 'manufacturing space' ................................................................. 159
- 7.8 Tool library specific tips ................................................................................. 160
- 7.9 G-code preamble .............................................................................................. 162
  - 7.9.1 G90 command ........................................................................................... 164
  - 7.9.2 G54 command ........................................................................................... 165
  - 7.9.3 G64 command ........................................................................................... 165
  - 7.9.4 G50 command ........................................................................................... 166
  - 7.9.5 G17 command ........................................................................................... 166
  - 7.9.6 G40 command ........................................................................................... 166
  - 7.9.7 G80 command ........................................................................................... 167
  - 7.9.8 G94 command ........................................................................................... 167
  - 7.9.9 G91.1 command ........................................................................................ 168
  - 7.9.10 G49 command ......................................................................................... 169
  - 7.9.11 G21 command ......................................................................................... 169
  - 7.9.12 G30 command ......................................................................................... 169
- 7.10 The operational G-code ................................................................................ 170
  - 7.10.1 N10 operational code ............................................................................. 170
  - 7.10.2 T7 G43 H7 M6 operational codes ......................................................... 170
  - 7.10.3 S2200 M3 M8 operational codes ........................................................... 171
  - 7.10.4 G54 command ......................................................................................... 171
  - 7.10.5 G0 command ........................................................................................... 171
  - 7.10.6 Commands to end a program ............................................................... 172

# Chapter 8    Useful PathPilot® G-code reference ............................ 175

- 8.1 G code ................................................................................................................ 175
- 8.2 M code ............................................................................................................... 176
- 8.3 Expanding on some finer points .................................................................... 177

- 8.3.1 G4 command .................................................................................. 177
- 8.3.2 M3 and M5 commands .................................................................. 179
- 8.3.3 M64 and M65 commands ............................................................... 179
- 8.3.4 G28 command ................................................................................ 179
- 8.3.5 G53 command ................................................................................ 180
- 8.3.6 G30 command ................................................................................ 181
- 8.3.7 G54 command ................................................................................ 181
- 8.3.8 G73 up to G89 canned cycles ........................................................ 182
- 8.3.9 M1 command ................................................................................. 183
- 8.3.10 Summary ...................................................................................... 183

## Chapter 9    Connection to Wi-Fi ........................................................ 185

- 9.1 Wi-Fi adapter on a USB stick, supplied with the Tormach® 770MX .............. 185
- 9.2 Mapped network drive .................................................................. 186
- 9.3 PathPilot® hub .............................................................................. 186
- 9.4 Dropbox .......................................................................................... 187

## Chapter 10    Tool holding ................................................................... 189

- 10.1 The basics of the BT30 spindle ................................................... 190
- 10.2 ER collets ...................................................................................... 193
- 10.3 Basic drill chuck ........................................................................... 195
- 10.4 End mill holders ........................................................................... 195
- 10.5 Face milling cutters ..................................................................... 196
- 10.6 Collet chuck for taps ................................................................... 198

## Chapter 11    Cutting tools .................................................................. 201

- 11.1 End mills, not slot mills! ............................................................. 201
- 11.2 Chamfer milling cutter / spot drill ............................................. 204
- 11.3 Face milling .................................................................................. 206
- 11.4 Engraving ..................................................................................... 207
  - 11.4.1 Engraving tool types ............................................................... 207
  - 11.4.2 Fusion 360 CAM engraving techniques ................................ 209
- 11.5 Machine taps ................................................................................ 211
- 11.6 Thread milling tools .................................................................... 212

## Chapter 12    Feeds and Speeds ......................................................... 213

- 12.1 The PMK chart ... 213
- 12.2 Looking up the catalogue data for the tool ... 214
- 12.3 Spindle speed in rpm ... 215
- 12.4 Feed rate ... 217
- 12.5 Depth of cut, Width of cut ... 218
- 12.6 Power considerations and Material Removal Rate (MRR) ... 219
- 12.7 Power calculation examples for the Tormach® 770MX ... 223
- 12.8 Torque considerations ... 225
- 12.9 Torque available ... 226
- 12.10 Torque required ... 232
- 12.11 What about the 10,000rpm option on the Tormach® 770MX? ... 232
- 12.12 Drilling holes ... 235

## Chapter 13  Work holding ... 239

- 13.1 Machine vice ... 239
  - 13.1.1 Vice soft jaws ... 240
- 13.2 Horizontally mounted 3-jaw chuck ... 241
- 13.3 microARC4 '4th axis' ... 242
- 13.4 ER40 collet chuck ... 242
- 13.5 A simple 'twin operation' work holding example ... 243
- 13.6 Clamping stock directly to the table ... 252
- 13.7 Use of 'tabs' for holding parts in place ... 253
- 13.8 Advanced fixturing techniques ... 253

## Chapter 14  microARC4 ... 255

- 14.1 Hardware in the kit ... 255
- 14.2 Connecting up the stepper driver control box ... 258
- 14.3 Keyboard control ... 259
- 14.4 Alignment of the A axis and centrality of the chuck ... 260
- 14.5 Fusion 360 and 4th axis work ... 261

## Chapter 15  Modifications and Additions ... 263

- 15.1 Passive Probe holster ... 263
- 15.2 Handheld safe air-blast gun ... 266

| | | |
|---|---|---|
| 15.3 | Hours meter | 268 |
| 15.4 | USB I/O M-code interface kit | 271 |
| 15.5 | Signal tower | 271 |
| 15.6 | Pneumatic solenoid for remote air blast | 275 |

## Chapter 16    Learn from Others ............................................................ 277

| | | |
|---|---|---|
| 16.1 | Lars Christensen | 277 |
| 16.2 | John Saunders | 278 |
| 16.3 | Mark Terryberry | 279 |
| 16.4 | Cliff Hall | 279 |
| 16.5 | Joe Pieczynski | 279 |
| 16.6 | Tormach® online | 280 |
| 16.7 | Fusion 360 | 280 |
| 16.8 | SkyCAD Electrical | 280 |
| 16.9 | Jayson Van Camp | 281 |
| 16.10 | Titan Gilroy | 281 |
| 16.11 | Norman Kowalczyk, Daniel Rogge, and Andrew Grevstad | 281 |
| 16.12 | John Grierson | 282 |
| 16.13 | John Ackerman | 282 |
| 16.14 | UK model engineering magazines | 283 |
| 16.15 | Autodesk Inc. | 283 |
| 16.16 | Joe Evans | 284 |
| 16.17 | Rob Joseph | 284 |

## Chapter 17 Formulae and G code summary ........................................ 285

## Chapter 18    Suppliers ........................................................................... 289

Notes.................................................................................................................. 294

# Chapter 1     Choosing the Tormach® 770MX

Starting at the very beginning, having been given the seed of an idea by my colleague Rob, it was time to properly investigate what was out there to decide on a suitable choice of machine. This book is about putting together and becoming conversant with the Tormach® CNC mill of the title, but I didn't start out knowing that was the one I would choose. I needed to do a little research.

## 1.1   The type of CNC controller

One of the first things you realise, when you begin to look at the range of hobbyist level CNC milling machines available, is they all need to be controlled in some way. By that I mean they must all have a computerised control system. And this is where you will find a sizeable difference in complexity between industrial machining centres and hobbyist/enthusiast 'domestic end-user' mills. And this is a good starting point to learn what's what. At this stage, bear in mind, I truly didn't know much at all about CNC machines. The following is what I found out.

    At the lower end of the scale, the truly 'hobbyist' style of CNC is often controlled by software named 'Mach 3'. I was aware of it but didn't know much about it. I found out it is also occasionally employed in some larger hobbyist equipment, and there is a later version called 'Mach 4.' I knew 'Mach 3' was originally based on older Windows computers when the 'parallel port' was still a thing. I had witnessed some of these combinations working at annual model engineering exhibitions, such as the one at Alexandra Palace in North London, and the equally impressive show at Harrogate in North Yorkshire (in the days before Covid-19 spoiled everything of course). Another common control system is an 'open source' suite of software perfected by enthusiasts running the Linux operating system, appropriately named Linux CNC (it was referred to as EMC2 in its early releases). At the other end of the scale in the manufacturing world, on machines costing hundreds of thousands, even millions of pounds, there are names such as Siemens, Fanuc, and Haas, and quite a few others which are industry-leading systems, way out of reach of most hobbyists, for sure. It occurs to me there are sadly few systems which blur the edges between the two groups. But one such system is PathPilot®, based originally on the Linux CNC building block, owned and developed by Tormach® who are based in the United States of America.

Tormach® began in 2001 with the specific intention to 'grey' the distinction between high end and hobby market, by designing machines which would be good for prototyping new products and equally could manufacture parts, albeit in relatively low volume, for emerging 'start-up' companies. The point being, lower size and cost had to mean lower speed, simply because you are forced to run at lower 'spindle motor' power levels of 1kW or 2kW, ten or twenty times less than industrial equipment, but this shouldn't mean their machines had to be lower quality. Tapping into the 'start-up' market demanded a decent repeatability and accuracy. It is true to say this market is more advanced in North America than in the UK or Europe. But hopefully not for too much longer.

Tormach® was certainly ticking my boxes and grabbing my attention. Naturally, for balance, you should look around at other machine manufacturers which are chasing the same market and, very quickly, you come across SYIL, a Chinese company. In my opinion, the products offered by Tormach® and SYIL are quite similar, and both companies are producing elegant machines for the market which continues to test the waters between industry and hobbyist. There are a few significant differences when you dig beneath the surface though. Tormach® is American, their engineering and design is based in Wisconsin, but the machines are manufactured in China and Taiwan (with some sub-assemblies completed in Mexico and USA). SYIL is Chinese based and Chinese owned. There are two aspects of each company's product range where the designs can differ. Firstly, some of the larger SYIL mills use linear guide rails in all three axes, which is the more modern way to do things. Tormach® are currently using the traditional dovetail slideways. Since these machines are not operating at industrial speeds, I am not convinced there is much difference in performance. Secondly, the SYIL machines come with options as to which CNC controller you wish to choose, with varying price tags depending on whether you go low end with Mach 3 or raise the bar with Fanuc or Siemens industrial controllers. (I believe some of the Siemens controllers will only be available to you if you purchase a 3-phase machine so, again, we are getting a bit industrial.)

The Tormach® does not offer any choice; if you purchase a Tormach® mill, it will inherently be a PathPilot® controlled machine. The question is, is it any good? Cleverly, Tormach® offer the ability to login to what they refer to as their PathPilot® HUB, set up a free account, and get to see how a virtual version of the control screen looks, long before you ever see the real thing. It is a very useful learning tool and you swiftly realise how user-friendly PathPilot® is.

All CNC controllers need G-code to work. How was I going to get G-code? Well, the previous year I had been introduced to desktop 3D printing and I was already getting familiar with 3D modelling in the DESIGN section of Fusion 360 drawing software, so it seemed an obvious and logical step to learn how to use the MANUFACTURE section within the same suite. Fusion 360 is one of the 3D modelling programs from Autodesk. Although Autodesk has been around for a long while, Fusion 360 is its young upstart born to challenge the dominance of industry leaders such as Solidworks. And it is very good indeed.

I had been designing parts for my desktop printer using the computer-aided design (CAD) part of Fusion 360 and found it to be intuitive. For personal use it was free of charge. However, I knew nothing about the computer-aided manufacturing (CAM) section of it. And the million-dollar question was, if I continued to use Fusion 360, would the CAM section of this software be compatible with my new machine?

When you delve deeper you realise each CAM software suite produces generic code since it does not know the brand of machine you will be using to make your parts. You need 'machine specific' code (each brand of CNC machines has its own idiosyncrasies and personalities) so you require some extra software, called a post processor, to interpret. The post processor is a piece of code which injects the personality of your specific machine manufacturer into your CAM software, so it always produces G-code which is compatible with your machine. This then begs the question: does Fusion 360 support post processing for the machine manufacturer you are considering? This is easy to check on the Autodesk website. And it is quite important, if you wish to use Fusion 360 CAD/CAM software, as I did.

A quick check online of the 'post library for Autodesk Fusion 360' reveals compatibility with both Tormach® and SYIL machines. In other words, Autodesk have written 'post processors' for both these machines. Interestingly, for some of the more advanced functions within the CAM section of Fusion 360, you must now pay a monthly fee, even if you are a personal user not making profits from its use. By the phrase 'advanced functions', they include automatic tool changer control and fourth axis work, so if you choose those options on your machine, you won't be able to produce code to control them unless you join their paid membership. It is an additional expense to budget for, but I found trying to imagine it as just another Sky TV package puts it into perspective. It really is excellent software.

After getting to this stage, I still hadn't decided which to buy. I re-visited the CNC controller aspect once more and began researching online for help, assistance, and advice on how to learn to use the machines. Be it from YouTube videos, online model engineer chat forums, CNC chat forums, or the companies themselves, it quickly became evident there was a whole pile of information out there for Tormach®, but not a great deal for SYIL. The difference is marked. What especially jumps out at you is the energy with which Tormach® themselves are promoting their machines by sharing information. Their openness and willingness to engage must be commended. The more information they can put out there, it seems to me, the more interest they will generate. That seems to be their ploy. It's refreshingly honest, and it certainly worked for me. After all, I was going to have to teach myself this whole subject and every little bit of help I could grasp was going to be invaluable. Tormach® are genuinely open to all discussion and their machines are always a work in progress. If you have an issue with some aspect of the machine, as happened to me, they will listen to you and take you seriously even though you are new to this. They want to help you master the machine. More importantly, they want to make their machines even better. I can't think of a better advertisement for a corporation than that.

Witnessing the online presence of Tormach® made my decision easy but, before I could sign on the dotted line, there was one other glaring aspect I needed to bring to the decision-making process. I was a home user in the United Kingdom with a 230 volts domestic power supply, which is single phase. I was never going to be able to consider a 3-phase 415 Volt supply in my home. I needed to be sure I was buying a machine which could run on a single-phase supply.

A quick check revealed all Tormach® and some SYIL machines still fitted the bill but, by now, my mind was already made up between the two.

The excellent PathPilot® control system, and Tormach®'s online presence and support, easily won the day for me personally. Now all I needed to do was decide which machine from their range, and which optional add-ons; not as simple as it might sound.

## 1.2 Choosing the Tormach® 770MX

The company website in the USA reveals they have a selection of milling machines of various sizes for sale, and the model numbers 440, 770, and 1100 represent three different levels of the sort of machine in which I was interested. Their website offers pdf downloads of detailed specifications for every model, a comparison spreadsheet, and the Operator's Manuals for each. I analysed these thoroughly.

The following are some of the comparisons you may wish to consider:

- Price
- Size of the overall milling machine
- Size of the parts it can make
- Electrical power supply requirement
- Stepper motors or Servo motors
- Tool holder type

### 1.2.1 Price

Price to pay and affordability are matters of a personal nature of course; everyone has this decision to make, but there are some aspects worthy of note. Firstly, if you are in the UK, you will want to know that Tormach® now have a preferred importer who will take your order and invoice you in UK pounds (GBP). This takes away some of the uncertainty of shipping costs, import duty and currency exchange rates. The company, at the time of writing, is CNC machine Tools Ltd, of Snetterton, Norfolk, UK.

Secondly, it is vital to look at the entire list of optional extras you may wish to consider and get the quotation itemised so you can see exactly where you could make savings if you need to. It is quite surprising how a headline basic price can rapidly increase when you start to specify many options for the machine. Tormach® are at pains to mention you can always add options later. But a few of those options would require some degree of dismantling to fit them retrospectively, the automatic tool changer an obvious example.

Thirdly, it is surprising how much money must be spent on cutting tools and tool holders. When a quotation comes from Tormach® for a new milling machine, it will not include any tools unless you have chosen to order some imperial sized ones from them. If you are not based in the USA, like me, you will most probably need metric ones. I buy all my cutting tools in the UK from high quality suppliers. Tooling is not cheap, but there is absolutely no point in buying such an accurate and excellent milling machine and then using sub-standard cutters. You must therefore budget for tooling separately.

### 1.2.2 Size of the overall milling machine

The size of mill you wish to purchase is a clear-cut decision. On the Tormach® website they provide pdf files with detailed engineering drawings of the layout of all their machines and what space they require around them, to be operated safely, and with the required access for maintenance and service. You must allow for this space – you will need access (more on this later) – so be sure to measure your available space carefully. The machines are delivered with the stand and mill separated and would normally be assembled onsite so, although you will need the published height available inside the building, you could accept an accessway with reduced height just to get the separate components indoors. Similarly, you will need some floor space in addition to where the machine will reside, to allow access for the hoist during assembly.

### 1.2.3 Size of the parts it can make

It is obviously the exact same parameters, and decision to be made, whether you are planning to order a manual or CNC type mill. Tormach® carefully specify each machine in terms of the available travel of the three axes, X, Y, and Z. They also specify the exact dimensions of the table in X and Y, which is obviously quite a bit larger. Keep in mind, you may wish to fit more than one type of fixture, such as two vices side by side. Also, if you are facing off a part with a large diameter tool, then you will need to 'clear' the part by at least the radius of the tool to do a neat job. It comes down to how big the parts are you think you might be making.

Also, you might want to make room for the electronic tool setting probe (see later) which cannot be located at the very ends of the table, because it must fall within the X-Y travel at least by its radius so the tools can reach it.

Finally, you will need to make allowance for which tool holders and tools you might be using as these take up some of the available space in the Z axis; again, this issue is no different to any mill you ever used.

### 1.2.4 Electrical power supply requirement

All Tormach® mills run from single-phase mains electricity. That's great news for the home user. The larger machine, the '1100', needs 230 volts. So, on paper, this is going to be the most suitable for the mains supplies in the UK and Europe. I appreciate we have 50 Hz mains frequency in the UK and the American machine is probably expecting 60 Hz, which is what they have in their country, but this is not going to be a factor of any concern. Firstly, the computer part of the machine works off its own switched mode power supply (SMPS) like any other computer, which you can run off just about any mains socket in any country you like. Secondly, the main spindle motor of the machine is powered by a special device called a Variable Frequency Drive (VFD) and controlled very accurately, independently of the mains supply frequency, by rectifying the supply to direct current and then creating its own 3-phase alternating current at a cyclic frequency of its own choosing. The remaining power supplies inside the electrical cabinet are similarly not going to be affected by a 10 Hz difference.

The smaller machines in the range, the 440 and 770, need 115 volts which we simply do not have in the UK. This is a small problem, but readily surmountable. I will cover this in detail in the chapter on preparations. In doing so, I will uncover a secondary problem which may be an issue with all three models of machine and I shall explain how to overcome this too. Don't be too concerned about any of this for the time being; if you have single-phase UK domestic mains, you can operate any Tormach®.

### 1.2.5 Stepper motors or Servo motors

The choice of which type of motors drive the three main axes of travel is entirely down to the buyer. Servo motors are fitted to the newer models and are more expensive. It is my understanding the smaller '440' mill always comes standard with stepper motor drives, but the larger machines have the option, which is reflected in their naming protocol. The Tormach® 770M and 1100M mills come with stepper motor drives, whereas the 770M+, 770MX, 1100M+ and 1100MX all feature servo motors.

There is plenty of debate on the internet about the two types of axis motion control, concerning achievable power, the possibility a stepper drive can 'slip' out of register if overloaded, and any differences in the speed and acceleration of motion. And I am sure, with careful design, both systems do a fine job.

But the thing which convinced me servo motors would be better, might seem light-hearted. Just take the time to watch some YouTube videos of people using stepper motor driven milling machines and turn up the audio when you do. The noise is distinctly 'stepper' harmonics and the machines almost 'sing' when any of the three axes are in motion. Depending on what part is being made, it will sing a different tune – circular motion is particularly recognisable. This rang true with me because I have a desktop 3D printer which has stepper motors and it sits on my desk at home and drives me crazy with its repetitive 'whining' noises. Unless you have heard a 3D printer doing this, it is very hard to appreciate how annoying it can get no matter how hard I try to explain. If you can afford the extra, I would recommend you choose servo motors.

1.2.6 Tool holder type

Early Tormach® machines had their own design of tool holder. They called it the TTS system. I think it stands for Tormach® Toolholding System, but I haven't seen that written down anywhere. I have never seen nor held a TTS tool holder but they are a common sighting on YouTube videos. The holders themselves are clamped in what is effectively an R8 collet, the type invented by Bridgeport many years previously. The '440' mill is fitted as standard with TTS. The latest Tormach® machines, however, utilise a slightly more rugged tool holder which fits an industry standard BT30 spindle, which is based on a taper and drawbar system.

The main aspects of debate between the two would suggest the BT30 is an improvement. It is more rigid. Also, the way in which the drawbar system works on the BT30 spindle is such that it tends to pull strongly upwards, with a better grip of the top of the pull-stud, resisting any tendency for the tool to be pulled out by cutting forces. A complication of this system is the BT30 spindle has two dogs on it which orientate and engage with the tool holder (as far as I know the drive torque transmission is still via the taper of the spindle, not the dogs). This means some sort of encoder system is required for positional control of the spindle, so the tool holder can be rotated into position for the dogs to engage correctly with an automatic tool changer. This added complexity, in turn, brings the advantageous capability of something called 'rigid' tapping, sometimes known as 'synchro' tapping. This is certainly worthy of consideration.

Finally, real benefit can be gained by owning several tool holders and pre-populating them with useful tools. Every time a new tool is fitted into a tool holder, it must be measured (more on how we do this later in the book) and the 'length data' loaded into the tool table of the machine's computer. By having a selection of known useful tools already measured in the system, it means much less time spent changing tools and re-measuring. The BT30 tool holders are an industry standard size so that might be a factor in persuading you to go down this route.

When I shared with Andrew Grevstad, Tormach's Business Development Director, my discussion about the pros and cons of the TTS system of Toolholding versus the BT30 system, he quite rightly corrected my partial misunderstanding. He explained the main advantage of the TTS system is the repeatable Z-height positioning on account of the fact it has parallel sides, slips into the R8 collet up to a known stop, and always engages this same reference surface against the spindle nose, even when making manual tool changes tightened to an unknown and variable degree. They market the TTS system as 'quick-change' on this basis. Conversely, the BT30 system engages on its conical taper only; strictly speaking, you should be re-checking the tool height after any manual change. If using an automatic tool changer, then repeatability is safeguarded, assuming of course the taper is kept clean and free of any debris.

Well, we covered all those bullet points above and I hope they help you as much as they helped me. Eventually, you must make an informed choice and every situation is different so I can't cover all possibilities and scenarios, but it's a start. I had made my decision to purchase the model known as a Tormach® 770MX. This has servo motors, needs 115 volts mains electricity, and uses the BT30 tool holding system. All I had to do next was decide what optional extras I wished to add to the base machine.

## 1.3 Purchase options

Choosing add-ons for the base machine is certainly fun, but it can also be the cause of an abrupt price increase. So inevitably this must always be a personal decision. Without extra add-ons, there is a perfectly capable machine to be had. But you will improve the capability with many of the options and you will certainly be increasing the convenience factor. Later, if you get into manufacturing parts in greater numbers then, of course, some options are going to make your life easier and, here's the point, save you time and money. Some of the extra 'bits and pieces' are just nice to have, and that's OK too.

I purchased most of the things which can be added to the 770MX. That's relevant because the rest of the book is going to be about installing those extras and learning to set up and operate them. I'll go through the list, which is directly from Tormach®'s website. Let's start with the largest and most costly add-on, the automatic tool changer.

### 1.3.1 Automatic Tool Changer (ATC)

For the Tormach® 770MX machine, the ATC unit comes ready built as one assembly in a large cardboard box marked 'made in Mexico.' That's a good advertisement; my 'Dremel' hand-held rotary tool was made in Mexico and has remained an amazing asset despite years of hard physical abuse in the workshop.

The automatic tool changer assembly comprises a carousel which takes up to ten tool holders (the ATC for the 1100MX takes twelve and for the smaller 440 takes eight) and is controlled by pneumatics and stepper motor technology. It is a heavy unit and, fitting it, you will most definitely need some help from a strong friend. It would be virtually impossible to fit after the outer steel enclosure of the whole mill has been assembled – you just couldn't gain the access needed. Taking the enclosure off would be a horrible task, not least because you will have to re-seal the whole machine against leaks. More of that later, of course, but the point I am trying to make remains clear; if you feel you will ever want an automatic tool changer, then try to buy it at the outset to save yourself a lot of extra hassle in the future.

**Automatic tool changer loaded with ten tools**

On the other hand, do you really need an automatic tool changer? Maybe not. Each time the program control calls for a tool change on a machine which has no automatic tool changer fitted, the machine will simply stop and ask the operator to make a manual change. It's easy, because whether you have an automatic tool changer fitted or you don't, the drawbar system (that's the collet system which grabs hold of the pull-stud on the top of each tool holder) is pneumatically controlled. You press a button; the tool holder drops out. You fetch a new tool holder, press the button again, and pop the new one back in. The only difference is you will get messy hands doing a manual tool change especially if you've been using coolant – the coolant gets right up inside the collets and drips all over the place. No big deal, but it does get everywhere.

I talk about installing the ATC later in the book. If you are undecided, I will leave you with this thought: it is simply awesome you can have a part being made with up to ten different tools being swapped fully automatically while you go and make a cup of tea. It really is very good indeed.

## 1.3.2  MicroARC4 – work holding in the 4th Axis

Another of the more costly optional add-ons is the 4th axis device they have named microARC4. This is essentially a stepper-motor driven 3-jaw chuck (also comes with collet holder option) which bolts onto the table and provides the ability to horizontally rotate the work while you are cutting it. The whole unit is quite heavy and requires careful planning as to how to secure it to the table. You'll need to source some tee nuts for it, as the ones Tormach® supply only fit the table of the smaller 440 series machine. I shall explain the details in the separate chapter on 4th axis work, and suggest you manufacture your own tee nuts to suit the table of your machine. You must also perform a minor modification inside the electrical cabinet, to install the stepper motor control box for this device. Even if you are not confident with electrics, this is not complicated and I will discuss that process too. The 4th axis unit is easy to add later and does not require any major dismantling of the base machine to fit it. If you need to keep the budget under control, then this is certainly an item you could put on hold. Unless, of course, you have an urgent 4th axis milling project already in the pipeline.

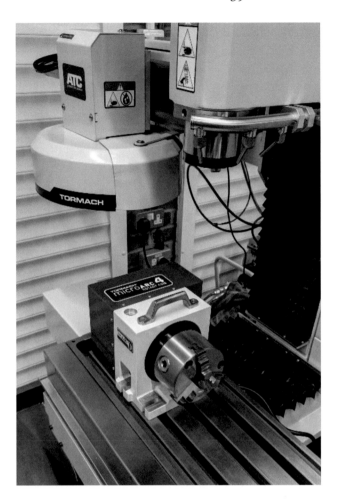

**microARC4 resting on the table**


1.3.3 PathPilot® Console

Now we come to the PathPilot® Console. On the latest 770 and 1100 machines, Tormach® have designed a professional-looking, well-made, heavy-duty framework into which they have installed a touch sensitive computer screen, with a tiny steel desk for keyboard and mouse. It is industrial in nature, tough and, to my mind, looks amazing. It really sets the machine off beautifully. But you do not have to buy this. The machine will work perfectly well connected to a more ordinary desktop computer (you supply the desk) and there is a small amount of money to be saved if you go down this avenue.

One thing to consider, in deciding whether you want to go the route of the PathPilot® Console is your available space. If you recall, I mentioned the Tormach® website has detailed drawings of all the machines' footprints and space required around them. In those diagrams, they show the console set at a 'trendy angle' so the screen faces slightly inwards towards the operator who is standing in front of the main sliding doors. It is fully adjustable and this looks smart, but in fact takes up more space than you'd think. To be truthful, it rather caught me out when I rotated it inwards towards me as I stood facing the doors of my machine, it used up too much space to get past the machine in comfort. So I decided to rotate my console screen back to parallel with the machine, so it faces squarely outwards. I don't think you could reasonably leave the adjustment loose, as you would compromise the light finger pressure on the touch screen control. I know some operators on YouTube imply leaving the whole display to swing loosely, but my machine did not seem to work this way, was too tight on the upright pole, and would have needed a lot of fettling to solve that issue.

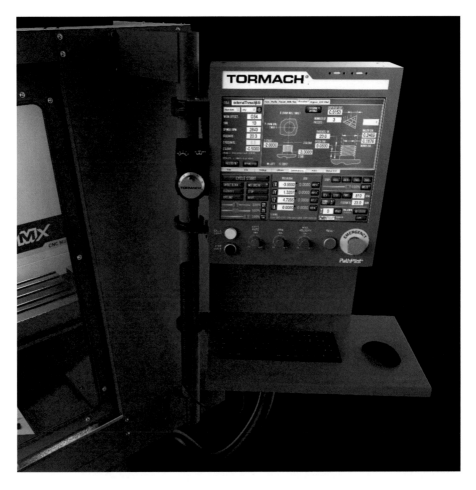

**PathPilot® console option – touchscreen operation**
Copyright© photograph courtesy of Tormach® Inc.


If you choose not to go with the PathPilot® Console, you will still need to set up some sort of desk and computer nearby, as already mentioned. This needs to be 'not too far away' as it must connect to the machine's electrical cabinet, so consider if you really are saving that much money or indeed space. You might make it worse for yourself. And you still need to buy a computer whichever way you look at it.

In my opinion the PathPilot® Console finishes off the machine superbly, making it look strikingly professional. It also functions well with the touch screen buttons, making it more intuitive than clicking on a screen with a mouse and a cursor, the latter not giving you the same tactile feel. It would be even better if they could incorporate some sort of 'haptic' vibrating feel within the screen assembly but that's for another discussion. The console also combines the cycle start, feed hold, emergency stop and power reset switches in a professional control panel you will be pleased to own. Yes, try to afford the console, it really is top notch.

1.3.4 Flood Coolant

Do you think you will need coolant? If you are planning to cut metal, then your response would almost certainly be yes. The coolant lubricates and cools, keeps the cutting tool sharp, and may improve the quality of the final finished part in some instances. It also washes away the swarf from the cutting edge. Swarf, or chips, getting between the cutter and the part can spoil the surface finish considerably and, in certain cases, break your cutting tools.

**Flood coolant in action**

Coolant is a bit messy, but the advantages are real. There are many other manufacturers' options online, such as mist systems, pure air blast systems, etc., and of course you can add any or all these later, but the main option being offered by Tormach® is what they call Flood Coolant. Comprising a 12 US gallon sump on castor wheels beneath the machine, a decent quality pump under the control of the machine itself, and a very smartly designed coolant delivery bar with three nozzles, this is certainly a credible option which works very well indeed.

### 1.3.5  Tramp Oil Skimmer

This was the first machine I had ever purchased which had a coolant system and, when you research the subject, you begin to appreciate it is quite a complex 'art' all its own. I realised I needed to take it seriously, without getting overly bogged down in the details. Looking after your tank of coolant is worth the time and effort. In the next chapter, about preparations, I shall go into some detail about coolant and its upkeep. But here, we are merely talking about choosing options for your machine. So, when you start to research machine coolant systems, you come across something called 'tramp oil.'

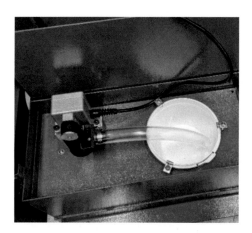

**Oil skimmer made by Skimpy supplied by Tormach®**

I want to keep this brief, trust me, we have many more interesting things to pursue, but tramp oil turns out to be an important consideration. Coolant is soluble oil mixed in suspension with water, the oil being specially designed for the job. Tramp oil, on the other hand, is oil which is not water soluble and shouldn't even be there, and so it floats on the top of the lovely milky white coolant in the sump. In doing so, it doesn't let the fresh coolant beneath the floating layer breathe properly and, over time, can make it go bad. Something to do with anaerobic bacteria, I believe. If it goes bad, it is unhealthy and it might smell bad too, so tramp oil is something we really want to remove as best we can. The coolant must be permitted to breathe. You may be wondering where this unwanted tramp oil originates. The parts of the machine which slide across each other, the 'ways', are lubricated by special oil called slideways lubricant and this can get washed into the main sump. It turns out a good way to remove this thin layer of unwanted oil is by using a small device with a motor and rubber belt which continuously pulls the oil out of the milky liquid below, and Tormach® refer to their tramp oil remover as an Oil Skimmer Kit. I recommend you put one of these on the order sheet.

### 1.3.6 Automatic 'Ways' Oiler

I briefly mentioned slideways lubricant above. It is special oil which clings to the moving surfaces of the machine so they maintain their slipperiness and every so often a little bit more is added by squirting into all the places it needs to be. This is accomplished via a gallery of tiny oil pipes which the manufacturer has carefully positioned around the machine.

On a hobbyist level machine such as the Tormach® 770MX, the choice is yours whether to buy a 'manual oiler' or an 'auto oiler'. The manual version comprises a small reservoir tank and a hand operated push-pull piston to manually give the machine galleries a quick squirt of lubricant every now and then, and this will normally be each day you intend to use the machine.

The alternative is buying Tormach®'s automatic version. Again, a small plastic reservoir which you fill with the special oil, and this time a motorised pump running at about 5psi which does the squirting for you. You can pre-set the electronic memory to deliver a pre-determined amount of squirt every so often, and it always gives one at first power up. It works very well and means it is something you don't have to worry about. Personally, I would recommend the automatic version every time.

**Automatic oiler supplied by Tormach®**

1.3.7  Electronic Tool Setter (ETS)

Later in this book, I shall describe how we set up the machine ready to start cutting. The setting up of any CNC machine is really the 'nuts and bolts' of the process and the machine requires to know two distinct facts before it can use the G-code to make your part. Namely, having secured a piece of material on the machine table or in a vice, the machine needs to know EXACTLY where that material is in three-dimensional space. Secondly, the machine uses a cutter in the rotating spindle and we must tell the machine's control system EXACTLY how long and how wide that tool is. These important items of information, without which the machine is unable to do its job, are called OFFSETS. In the simplest terms possible, if the machine knows where the work is, and how big the cutting tool is, then it can bring the two together and make your part.

Let's consider tool offsets first. We can measure tool offsets manually or with a probe and the Electronic Tool Setter is one such probe.

The tool offset data for each tool must be measured and entered in a tool table, rather like a spreadsheet, in the machine's controller software. So, for example, if we had an 8mm diameter milling cutter and tightened it into a tool holder, we would need to assign three parameters to that tool/tool holder combination. Firstly, we must assign a notional tool number, for example tool #23 so that, all tools being uniquely numbered in the machine's list, it won't get muddled up. It can be any number you like, except '99' which is reserved. Secondly, we know its diameter is 8.00mm, so we type that into the tool table. Thirdly, we need to know the total length of the tool/tool holder combination. This parameter is the trickier one to measure and there are many ways it can be done. When you think about it, all we are trying to do is find a way of referencing the length of all the tools in the list, so that if we tell the controller to use a particular tool, the machine will refer to the tool table, find the data it needs and then move the tool downwards towards the workpiece inherently 'knowing' when it is about to make contact. But the question will dawn on you, when we measure tool lengths, whilst we understand we are measuring down *to* the tip of the tool, where exactly are we measuring *from*? Whichever way we choose to reference this measurement, that question gets answered by following the procedures laid down for each of the various ways of doing it (and there are quite a few). It's my position the electronic means of doing it makes it so easy I can't think of any reason why I would ever wish to revert to a more manual method. By 'electronic' means, I am referring to the Electronic Tool Setter (ETS) device which Tormach® offers. This is not an inexpensive item, but it is an extraordinarily easy item to use, and I believe it is most certainly worth your serious consideration.

**ETS probe secured to the table**

Just remember whenever you change a tool in a tool holder, you need to measure it again. Even if you are putting the same tool back into the same holder, it is impossible to be sure you tightened the tool up in the same way as last time so it needs to be re-measured and entered in the tool table.

The wonderful advantage of the ETS system is you can bolt the device on the machine table, tell the machine software where it is located (horizontally in X and Y axes) and then, whenever you wish to measure a new tool, you simply press a button (a pseudo-button, it's actually on the touch screen) and the machine takes the tool to the ETS probe, senses it, and enters the tool length data into your tool table for you. It is very accurate, very fast and fully automatic. What makes this process quick is the fact the probe is working in conjunction with software already written as a macro within the PathPilot® architecture, making the process effortless and error free. There is a comprehensive section in the Tormach® Owner's manual which describes how this ETS probe is initially set up – of course, at the outset, we need to teach the machine the 'Z' reference to which we are referencing the probe itself. It is fiddly but you only need to do it the one time.

Once again, it is important to realise this is an option. There are other ways to measure your tool heights, more labour intensive, slower, but certainly cheaper too. Tormach® offer a Tool Height Setter, which is basically a dial gauge built into a metal housing with a big return spring and this is a good and cheap way to go if you cannot justify the electronic probe. And, of course, you can always buy one later if you wish to curtail your initial budget. Would I recommend an electronic probe though? Absolutely I would.

### 1.3.8 Work Offset Probe

In Tormach® 'speak', the Passive Probe is a device for measuring work offsets. It is electronic and fits in its own tool holder. It has a cable attached to it, and this cable plugs into the electrical cabinet of the machine. It is relatively inexpensive and works well.

**Passive Probe in author's own design of holster**

Just like the ETS probe for calculating tool offsets, this device also works under the control of a special macro which Tormach® have included within their PathPilot® software. Not only are you able to probe where your work is in all three dimensions, but you can also do clever things like find the centre of a hole, or a boss, for example.

Just like tool offsets are entered automatically into the tool table, the work offsets – three values of X, Y and Z - are individually entered into the work offset table for you. This whole subject gets considerably more complex when you start to talk about the Work Coordinate System of a CNC machine and I shall go into this in much greater detail later. But, for now, it is important to realise that you will need to have at least one way to measure this. You simply cannot operate a CNC machine without having the means to be able to tell your machine accurately where your work is located.

However, there are various options. There is a different electronic probe called an Active Probe. I do not know much about this option other than it is vastly more expensive than the Passive Probe, so presumably it has some clever proximity detection system rather than merely gold-plated switch contacts. The specification data suggests it is quite a lot more accurate with repeatability of 0.002mm compared with 0.015mm. It is four times the price so I guess you must make your individual choice. But there is another way.

The alternative to using electronic work offset probes is the use of a mechanical device. Basically, these are dial gauges which you 'zero' in to find an edge. So, they are edge finders, but cleverly they can detect in any lateral direction (hence they can find an edge in the X or Y axes), and they can detect in the vertical (thus are able to find where your workpiece begins in the Z axis). Much like a vacuum cleaner became known as a 'Hoover', these three-dimensional edge finders have become synonymous with the name 'Haimer', the German company which invented them. These are quite expensive tools since they are highly accurate mechanical devices. Certainly, worthy of consideration and, of course, there are cheaper copies on the market.

The mechanical edge-finder type probes require manual entry of the data into the work offset table whereas, as mentioned previously, with the electronic probes this is automatic. If you don't trust yourself transferring numbers reliably and error-free that might be a factor in your decision. I decided to add the Passive Probe to my shopping list and in a later chapter I shall talk through operating this device on the Tormach®. It works very well indeed.

**Tormach®'s passive probe attached to BT30 holder**
Copyright© photograph courtesy of Tormach® Inc.

## 1.3.9 Work holding

The subject of work holding is potentially huge and I cannot pretend to cover all the aspects of it in this book. Sometimes you will be clamping your material to be machined directly to the machine table. Other times you will want to clamp your work in a vice, which is itself bolted to the table. As you can imagine, there is an unlimited and constantly expanding range of products to serve these needs. Particularly when you wish to manufacture multiple instances of the same part in the one 'sitting', this subject quickly gets very involved.

But you must start somewhere. I ordered two separate work holding solutions straight away, a Tormach® vice and a Tormach® clamping kit.

The vice comes with bolts and tee nuts for securing it to the machine table. The table has three tee shaped slots, each 5/8-inch width, and the set is compatible with this. So, buying a vice from Tormach® made sense to me. I chose their 5-inch standard machine vice. Our colleagues in North America spell it 'vise' but we will forgive them that.

The clamping kit, again compatible with the 5/8-inch table slot size, has a selection of tee nuts, bolts, top nuts, and clamp bars so it is possible to clamp just about any shape and size of work directly to the table. This is the same technology as you would find on a manual milling machine so really no surprises here.

Likewise, just as when using a vice on a manual mill, you will likely support your work in the same manner in a CNC mill, so it projects above the vice jaws at the height you need. To do this accurately you will require a set of parallels, finely ground hardened metal bars of varying thickness made for the purpose. Order a set of these too if you don't already own one.

For more advanced 'work holding' solutions, there are some beautifully engineered solutions based on bolting a fixture plate to the machine table and then using special 'biting' tools secured to this plate to hold the work. I can wholeheartedly recommend you make an online search for John Saunders, where you will find Saunders Machine Works and his fixture plates, and NYC CNC, his YouTube channel, for all things CNC. The man is a legend in his field and deserves a mention. He is also very pro-Tormach® and an expert with Fusion 360 software.

### 1.3.10 Tool holding

As already explained, the Tormach® 770MX edition already comes fitted as standard with a spindle which only accepts the BT30 type of tool holder. Consequently, you are going to have to purchase some BT30 tool holders. There is no choice in this matter: they are 'optional extras' in that they do not come included, but if you don't buy any, you aren't going to have much success with your machine.

The question soon arises, which tools are you going to be using and therefore which tool holders will you need? This is not a simple thing to address because I do not know what you plan to make with your machine. However, to get things moving, Tormach® have thought of this and I recommend you consider ordering their BT30 Operator's Kit, which is a selection of tool holders to get you started.

I came across some confusion between metric and imperial kits on offer from Tormach® and I was initially supplied an imperial kit, having placed an order for the metric version. Whilst Tormach® insist they will supply you whichever you want, at the time of writing the metric version still does not show up on their web page, so hopefully this will have been resolved by the time you read this. It is an important consideration though because, generally in the UK and Europe, you will be using metric cutting tools. The matter was resolved very easily by the UK preferred importer, CNC Machine Tools Ltd., simply replacing those parts of the kit which needed to be metric, locally sourced in the UK.

I recommend the starter kit wholeheartedly but of course you can source all the tooling and tool holding you need in the UK simply because the BT30 spindle size is an industry standard size, albeit on the smallish side. I devote a whole chapter to tools and tool holding and it becomes a fascinating subject, the more you investigate it. Some of the major tool supply companies, certainly in the UK, will send you hugely comprehensive tool catalogues in the post, if you ask nicely, from which you can learn a great deal.

Personally, I think the Tormach® BT30 ER20 collet holders are excellent and I ordered a few more of those, adding to the six which came in the kit. I also ordered a few more of the regular sizes of collet I knew would be useful, such as 6mm and 8mm, simply because those sized end mills are common and very suitable for the Tormach®. You don't need to buy any cutting tools from Tormach®; they really are devoted to the imperial 'inch' system of measurement, and you are best off buying all your tooling in the UK (or Europe of course) if you are planning to operate your machine in metric.

### 1.3.11  USB M-Code I/O Interface Kit

This is an interesting option. It is comparatively inexpensive (a couple of hundred dollars or thereabouts) so whilst I had a very large consignment on order from the United States, it seemed to me a no-brainer to order one of these at the same time. I did not know what I would use it for at this stage but I figured it would be useful sooner rather than later. Simply put, this is a small USB connected 'input/output' device which has four inputs (to detect binary states, such as switch inputs) and four outputs (relay switches which can control other circuits). The G-code which controls the machine has some extra commands which may be added to your programs to consider the inputs or command the outputs. In a later part of the book, I shall explain what I chose to do with it. You may very well already have something in mind I hadn't thought of.

### 1.3.12  Lifting bar kit

Tormach® machines must be assembled at your premises and one particularly important stage of this is the lifting of the milling machine itself onto the base or stand. For this you will need access to a hoist, which is discussed in the next chapter, but additionally you will need to purchase the Tormach® Lifting Bar Kit which is designed specifically to lift your machine safely and in balance. It is listed as an option but you pretty much must purchase this – otherwise I cannot think how you would be able to proceed. Even if you had access to some strong nylon straps and a forklift truck, you would still find it difficult to suspend the milling machine exactly horizontally to place it accurately down onto the stand. This kit does this extremely well and safely. You really need this option.

I am advised by Tormach® there are holes in the bottom of the casting of each of their mills to allow for lifting bars to be slotted through, giving a different means of lifting the mill onto the stand, from lower down, rather than from above, but this would need specialist equipment to manage properly and safely and is probably not so useful for the hobbyist installer.

# Chapter 2    Machine specification

Allow me to summarise the Tormach® 770MX milling machine in simple terms to get a feel for its size and ability. I'll be completely honest: I am only repeating data which I have taken from their website but I am sure they won't mind me passing it on.

Copyright© photograph courtesy of Tormach® Inc.

I stand 6'2" tall (1.87m) and the Tormach® 770MX machine is as tall as me so that immediately conveys the size. Anyone with an appreciation for engineering, I am sure, will agree the photograph shows it to be a fine-looking machine.

So, whilst on the one hand it could be argued this is a hobbyist level, prototyping, and model engineering product, it doesn't take much imagination to see this machine in a small production environment for a start-up organisation eager to dip its toe into manufacturing. Tormach® are blurring the boundaries with this machine and it's wonderful to have access to such accurate machining for the price.


## 2.1   Tormach® 770MX specifications in detail

The Tormach® 770MX is a CNC milling machine with its own control software which is called PathPilot®. It is normally a 3-axis machine but has the capability to run a 4th axis simultaneously if you purchase the option separately. At this present time, I do not believe PathPilot® is equipped to run a fifth axis. PathPilot® is compatible with the use of professional CAD/CAM systems such as Autodesk's Fusion 360, which specifically offers a post processor for Tormach® mills. That means it is relatively straightforward to create complicated G-code to manufacture advanced parts. The 770MX has servo motors instead of stepper motors so it runs much quieter than traditional hobbyist level machines. The PathPilot® controller also has 'conversational' modes for ad hoc milling operations where full computer-aided manufacturing has not been necessary. With the addition of the automatic tool changer option, the 770MX becomes a machining centre capable of making small parts fully automatically. It has a relatively small footprint and runs from a single-phase mains supply of 115V making it suitable for the domestic market. In the UK and Europe, however, this will tend to require the need for a mains step-down transformer.

Let's have a look at the figures in greater detail. The machine is not especially powerful. In subtractive manufacturing – removing the material you don't want – there is a limit to the rate of material removal per watt of power you put into the cutting operation. The specific energy required for material removal rates (how many joules of energy you need to remove a unit volume of material) depends on how tough the stuff is to cut! The rate at which you can do it depends on the power you have available. Industrial machines might typically have spindle motor power ratings in their tens of kilowatts. Clearly, then, the

Tormach® 770MX with a 1.5HP spindle power rating isn't going to be a troubling competitor. (1.5HP is equivalent to 1.12kW). But that doesn't mean to say it can't do the job – merely that it will take a little longer. If you put the energy in, you'll get the job done. If you are running at a lower power level, this simply means you are providing the energy to do the work at a lower rate. Slower.

That is why Tormach® may confidently state their machine can cut stainless steel and even titanium. They are not saying it can plough through these like butter – it's going to take a little longer. But the main point is the machine is rigid enough to be able to offer the cutting tool up to these tough materials and successfully cut them without chatter and vibration. This is excellent news indeed.


2.2   Feeds and speeds – first thoughts

When you first start cutting metals with the 770MX you soon get an idea of how the 'feeds and speeds' are expressed. In other words, the movement of the cutter in any axis toward the material (feeds) and the rotational rpm of the cutting tool (speeds). In the UK and Europe, we tend to use mm/minute for feeds whereas in North America they will be using IPM (inches/minute).

Tormach® quote maximum feed rates in the X and Y directions (that's the table moving left, right, forward, and aft) of 300 IPM. That's 5 inches per second! It is impressive the axes can move that quickly to get to where they need to be to begin a cut, but with a spindle power of only 1.12kW do not be persuaded you will be cutting at these rates.

It might help us in the UK to try to visualise the metric feed rate figure. We will be running our machine in the metric system so all rates will be quoted on the PathPilot® controller screen in mm/minute. I find it difficult to visually gauge the 'feed rate' in my mind's eye from numbers expressed in these units. They are a bit meaningless initially. To help, I divide by sixty since I can easily imagine the passing of one second and better conceptualise the movement of the cutter in that time. Say for instance, I choose a feed rate of 480mm/minute. Divide it by 60. That gives a feed rate of 8mm/second. Yes, that's much easier to visualise, don't you think?

## 2.3 Spindle belt drive settings

The spindle of the 770MX has a belt driven pulley arrangement with two fixed options. Tormach® quote maximum spindle speed of 10,000rpm. If you place the belt on the lower setting, then the maximum is 3,250rpm. I have been using both belt settings with excellent results. There are considerations worthy of note when deciding which setting is more appropriate for the job in hand. By using the lower gearing, you are getting greater torque which may be important for tougher materials. However, the Tormach® is powerful enough to operate at the higher speed, especially for materials like aluminium alloys. And in fact, it is much quieter operating it at 5,000rpm on the high ratio than operating it at 3,250rpm on the low ratio, since the motor is going half the speed.

## 2.4 Table size and machining volume in x, y and z

The machine table of the 770MX is 26 inches x 8 inches, that's 660mm x 203mm. It has three parallel slots which are 5/8-inch wide and these accept tee-nuts for securing work directly to the table or clamping bolts for a vice.

The table cannot move as far as that though; its maximum travel beneath the spindle head is 14 inches (356mm) in the X axis (that's left and right) and 7.5 inches (191mm) in the Y axis (that's forward and aft). It's perfectly normal the table motion in the X and Y directions is less than its dimensions.

The Z axis (up and down) maximum travel is 13.25 inches (337mm) and that sounds a lot, but don't forget you need to put a BT30 tool holder into the spindle, and the BT30 must have a cutting tool inserted too. Then you might be holding your work in a vice which is bolted to the table. All this uses up space in the Z axis so the amount of travel available to make any parts in the vertical direction is going to be appreciably less than this headline figure.

## 2.5 Synchro tapping and thread milling

The spindle motor control is quite advanced on the 770MX. It has positional control as well as speed control. It needs to have positional control because the BT30 tool holders have dogs which engage with the spindle drive, so each tool must be stored in the automatic tool changer by first orientating these dogs to engage with the carousel. When the tool is retrieved, the spindle must reorientate itself so the tool is properly aligned again before the spindle can accept it. It is a sizeable advantage to be able to use the BT30 industry-standard tool holders, which are a bit chunkier than the original Tormach® TTS in-house design.

But, as an added bonus, since the machine has this positional control facility, it lends itself to a clever thread forming process called rigid tapping. Sometimes referred to as synchro tapping, the tap is driven into a pre-drilled hole and then reversed out while the height of the tool (the Z axis control) is controlled to very fine tolerances so the thread is not compromised. In other words, there is a direct correlation between the movement of the tapping tool in the vertical direction and its rotation in the spindle. This all happens extremely quickly and looks to the human eye as if the spindle reverses back out of the hole almost instantaneously. Rigid tapping makes fast and reliable work of threading holes.

For cutting larger diameter threads, in which the forces would be too great for tapping, the 770MX can perform 'thread machining' whereby a thread forming milling cutter tool takes a helical path down and around a thread, and it can do this for both internal and external requirements.

## 2.6 Coolant system

The 770MX is set up for flood coolant and has a decent sized sump with about 12 US gallons of coolant which is pumped to nozzles bolted on the side of the milling head. This is all switched on and off by the PathPilot® controller software, and consequently is also under the control of your chosen CAM software.

## 2.7 PathPilot® console controller

The latest PathPilot® console has a professional looking touchscreen, robust steel 'mini' desk for computer keyboard and mouse, and benefits from full Wi-Fi connectivity. All in all, the latest 'MX' versions of Tormach®'s mills exhibit impressive capability for their price. If you decide to place an order for a Tormach® machine, you will need to make some preparations before it arrives.

# Chapter 3  Preparations for delivery

## 3.1  Allow enough space

By now you know my personal choice was to buy a Tormach® 770MX milling machine. I will talk about this mill specifically since this was my experience. If you have a different machine on order, you will appreciate the same considerations will need to be made regarding your own preparations before you take delivery and you will need to do some careful research.

Tormach® publish online some nicely drawn 'Machine Footprint' diagrams for all their machines. I have included the drawing for the 770M machine (they do not appear to have caught up with their nomenclature yet but as the MX version is the same basic machine as the M version, just with some fancier servo motors, I figured this drawing would suffice. It did).

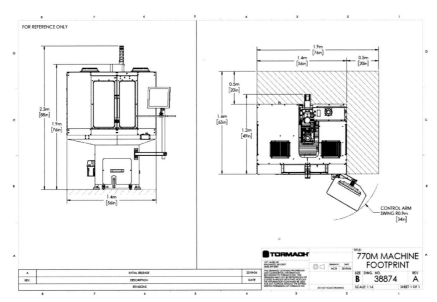

**Machine Footprint Diagram**
Copyright© image courtesy of Tormach® Inc.

In truth, this is a drawing of the machine with a standard computer screen on the front, and not the newer PathPilot® Console option. They haven't caught up with that drawing either at the time of writing this. But it doesn't matter too much; you get the picture, and the sizes will be very similar. I will never criticise Tormach® for this lack of being up to date with the latest drawings; they are producing some fine machinery at their chosen price point, they are employee-owned, and I am pretty sure working their socks off continually developing newer and improved versions. Would I prefer average equipment, but the paperwork the most fantastic ever and always up to date, or leading-edge machining centres (for the hobbyist end user market) but paperwork occasionally found wanting? The answer is surely obvious. Keep doing what you are doing, Tormach®!

Take a look at the blue hatched area in the diagram. That is not the machine's footprint, but the space they suggest you must keep free of obstructions around the back and side of the machine for suitable access. And now I have my very own machine in my garage, I can see why this is important. Don't be tempted to think this is optional, you really will need access to these areas for both the assembly and future maintenance of the machine.

Well before you even place your order for a machine, research your space availability and, certainly once the machine is on its way, make some detailed plans for exactly where it will reside, even going so far as to mark the floor with a felt tip pen. You will notice the controller's computer screen is set off at an angle in the drawing but this is just an option in reality; you can tighten the screen assembly in any position you like. In fact, you would think you can leave the screws on the loose side so the screen can be swung in and out to suit your chosen viewing angle while you operate the machine but, in my experience - and don't forget I chose the 'premium' PathPilot® Console – it really is not designed this way. If you leave the screws loose so it can rotate on the vertical pole, they just work looser still. So, they really need to be 'nipped up' tight, and then the whole assembly is rigidly clamped. You choose a screen angle and stick with it. What I had not quite appreciated was the extent to which the screen at an angle uses up space so I had to set my screen back in line with the front of the machine, so I could still walk past. It still looks very professional.

## 3.2 Download the Operator's manual (pdf)

The Operator's manual for every Tormach® seems to be a combination of Installation Procedures and Operating Instructions. These are available free online on the https://Tormach®.com/ website. In fact, when the machine is delivered, somewhere in one of the palletised boxes will be a copy of the manual in a shiny plastic ring binder, as I hope you would expect. Nevertheless, I think it is useful to get a copy early and start to think about some of the aspects of this installation which will need some planning ahead.

Download a copy onto an iPad or get one printed for a few pounds (that's what I did, it's personal preference but I like to make pencil notes in the margin). I have the Tormach® 770MX Operator's manual (version 0720A) and chapter three is titled 'site requirements', which is clearly a relevant and necessary read. The site requirements, aside from the physical space required to accommodate the machine, tell you what electrical power supply and what pneumatic air supply is required to make it work.

## 3.3 Subscribe to Tormach®'s YouTube channel

One of the selling points for me, when I chose to purchase a Tormach® CNC machine was their extensive use of YouTube to help users in all aspects of their machines from installation through normal everyday use to maintenance problem solving. You are going to have to assemble this heavy machine and, in the wrong hands, this could prove dangerous. You will want to watch how it is done on their video. There simply is no better way to explain the 'how to' to someone than to show them. Tormach® do this extremely well. Subscribe to their channel and start to get a feel for what is headed your way.

It is at this time you will begin to realise there are a few 'big hitters' out there on YouTube – influencers who have in recent years become known as brand ambassadors – who are promoting Tormach® and telling their individual stories of how they started out with their first Tormach® and developed businesses from there. Mostly in the United States, it is fair to say but certainly not exclusively, these videos make for fascinating viewing and will absolutely and undoubtedly enhance your understanding of the machine.

## 3.4 Electrical supply

This is an interesting subject. But I would say that because I like 'electrics.' I could just make the statement: *the instruction manual says you need a 115-volt single-phase supply (50Hz or 60Hz) on a dedicated circuit breaker of 15-amp rating.* And then I would suggest you phone up your local electrician and ask them to sort this out for you. But that could get expensive. Putting a dedicated mains supply into a room, or a garage, may not be that simple for many people. It was not a realistic option for me. My garage is the opposite side of the house to where the mains distribution board is located – it would have meant pulling up carpets and floorboards to achieve this.

I don't think it's necessary. There is more to the electrical supply requirements than you might initially suspect, and it is worth spending some time looking at it constructively. With some better understanding, I am hoping I can ease some of the pain.

3.4.1 Dedicated supply?

First things first, taking data from the operator's manual, a quick 'back of a cigarette packet' calculation reveals if the machine did take 15 amps at 115 volts, it would be operating at 1725W (watts). That's highly unlikely, as it's the intended circuit breaker's absolute limit, so we should expect it to be drawing slightly lower current than this, naturally. The machine's biggest power draw is the main spindle motor (1.1kW according to the manual specification) and the three servo motors which move the machine's three axes. Let's say, for the sake of discussion, the machine is likely a 1.5kW machine which could peak intermittently at up to 1.7kW.

Now, the electric kettle in my kitchen is rated at up to 3kW and I happily plug it into a socket on the wall, which in turn is wired into my domestic ring main connected to a 32A (amp) circuit breaker. My kettle manufacturer didn't tell me I had to use a dedicated supply so why are Tormach® making such a big song and dance about their machine, which operates at about half the power?

Delve a little deeper and nestled in the back pages of the Tormach® 770MX Operator's manual you will find the electrical schematic diagrams for the machine. Amongst those, you will find that inside the electrical cabinet there are four miniature circuit breakers (MCBs) labelled CB1, CB2, CB3 and CB4. The machine has been designed with ample protection built in. CB1 is the 'main breaker' rated at 15A. The remaining three breakers are all connected through this one; CB2 is for the machine motors etc. and is also rated at 15A, CB3 is for the accessories such as the coolant pump and rated at 10A and finally CB4 is for the auxiliary circuit (basically just the enclosure floodlights) and this is rated at 3A. Note that all power for the entire machine must go through CB1.

The Tormach® 770MX is a well-designed, well-protected machine which comes with a plug. Unsurprisingly it is a North American standard 3-pin mains plug called a NEMA 5-15P (which means a plug rated at 15A with two poles and an Earth pin). The NEMA plug is very different to UK or European plugs so it would be impossible to accidentally connect it to the 230V mains in the garage, for instance. When bad things are impossible it's a good thing, so I believe it is vital to leave the American plug untouched, forcing us to figure out how we are going to provide a compatible wall socket later.

In terms of the special requirement on voltage, yes, this is a dedicated supply in that it's a different voltage to normal for the UK and Europe, but I don't think that is what Tormach® meant. The implication is for a separately fused supply, direct from the mains distribution board, to power the Tormach® alone. If that is possible in your situation then that's fantastic, but many garages, certainly in the UK, will have one lighting circuit and one mains supply circuit. In my case, the latter is a 32-amp ring main.

Based on total power demand alone, and forgetting the voltage problem for the moment, you would not have expected this machine needed its own dedicated supply. If an electric kettle can plug into the ring main, then why shouldn't the Tormach® be permitted? It is my belief there is another reason Tormach® have specified a requirement for a dedicated supply. It's possible they are trying to address two ancillary issues; electrical noise suppression and something called ground fault circuit interruption, rather than general concerns about absolute power requirements. Their 'Site Requirements' (electrical power requirements) chapter goes on to make these additional points: (Copyright© Tormach® Operator's manual)

*"Primary power must be provided by a dedicated circuit, which must be isolated from electrically-noisy devices like welders or plasma torches. The machine should be isolated from inductive loads like vacuum cleaners or air compressors"*.

*"Power for the machine must not be protected by a ground fault circuit interrupter (GFCI), as it interferes with the operation of the variable frequency drive (VFD) spindle controller"*.

*"You must properly ground the power input to the machine"*.

Let's address each of these three fundamental statements in turn. Hopefully I can convince you there is a solution whereby you can operate safely and reliably from your garage ring main. But it is going to take a bit of thought and careful explanation.

Firstly, welders and plasma torches take a lot of power so I suggest you might wish to 'schedule' the use of your machines. We are not talking about an industrial scene here. This is a home shop environment on a domestic mains supply. You can do some welding, do some plasma cutting, run your bandsaw, power up your shiny new CNC milling machine, or even do some actual vacuuming in the garage. But please, not all at the same time! Your 32A ring main is not designed to do all this stuff at once, so be sensible. Be realistic.

That leaves the elephant in the room, the air compressor. You absolutely must have an air compressor because this Tormach® mill needs a pneumatic supply to function. And its pump will 'cut in' seemingly at random, when it must 'top up' the reservoir – you don't have any control over when it does this - so we had better make sure we can run our air compressor without fear of the electrical noise it generates causing havoc with the computers inside the Tormach®. All air compressors have powerful motors which generate electrical noise. It is a problem which must be solved. I fitted a snubber circuit to my compressor, a component from RS Components Ltd., (their reference RS813-430), and you can read about how I did this in the section on compressors.

Let us now attempt to address the second point. Well, first off, let's do a little 'across the North Atlantic' translation. The Americans speak about Ground Fault Circuit Interrupters (GFCI) and in the UK we prefer to use the term Residual Current Devices (RCD), Residual Current Circuit Breaker (RCCB) or even Residual Current Breaker with Overload Protection (RCBO). Not so long ago, we also used the term Earth Leakage Circuit Breakers (ELCB) but this has taken a back seat. All these terms refer to a special sort of circuit breaker which detects when the electrical current going 'to' an electrical load (anything you have plugged in) is different to the current coming back 'from' it. Typically, in the UK, this detectable difference is designed to be as sensitive as 30mA (milliamps or thousandths of an amp) and the RCD device is designed to instantly switch off the power if this happens. This is a safety feature to protect us all from electrocution. Modern regulations require the use of these RCD devices in our domestic mains distribution board. Sometimes you will see them added to extension leads for things like electric lawn mowers.

The problem comes when we have a machine which uses a Variable Frequency Drive (VFD) to control the speed of the main spindle motor, something the Tormach® does indeed have. Variable Frequency Drives operate at high frequency and consequently generate a lot of electrical noise, interference all-around as radio energy called Electromagnetic Interference (EMI), and interference existing as unwanted tiny currents in the cables called Radio Frequency Interference (RFI). This EMI/RFI plays havoc with computer systems and so the VFD device must have a special filter added to reduce the interference down to acceptable levels. This is termed an EMI filter and, typically, it is an electrical component consisting of capacitors and resistors connected across the mains input to the VFD and, crucially, also from each of the live and neutral wires to Earth (or Ground as the Americans refer to it).

Why should this present a problem? Simply put, the unwanted electrical interference is 'soaked away' to Earth by the action of the EMI filter. However, recall in a normal domestic electrical supply there is an RCD device whose sole purpose is to look out for any 'leakage' and it is unable to differentiate between intended 'soak away' to Earth of unwanted stray signals and genuine fault currents to Earth which were not intended. So, it does its job and correctly disconnects the power, even though of course we don't want it to. The two devices, the EMI filter and the RCD circuit breaker are, in this combination, incompatible.

We have a problem. We certainly do not want to remove the RCD breaker protection from the 32A ring main in our garage. That would not be a safe thing to do, especially in a garage environment. And it would not be in keeping with the regulations. All the other tools in the garage which run on mains electricity we wish to continue to be protected by the RCD breaker for our own safety. But we accept we are going to have to run the Tormach® machine on an electrical supply which is not protected by this feature due to the specific nature and demands of this complex and special case.

There are two reasons, therefore, I believe Tormach® are advising the machine needs to be connected to a dedicated supply. The supply must *not* have Residual Current protection. And being on its own separate supply will mean the Tormach® is more electrically isolated from any electrical interference generated by other loads nearby, such as the air compressor, which would be plugged into a different supply. But if we don't have a separate dedicated supply available to us, only a single 32A ring main, is there a suitable solution?

At this point, it is time to solve the voltage 'problem'. Because, happily, it turns out that having to step our UK (or European) 230V mains voltage down to 115V solves the issue quite neatly. We are going to have to invest in a device called a step-down transformer. By using a transformer, we will be able to run all our other electrical items on the garage ring main with full RCD protection and yet run the Tormach® without any RCD protection, which is what we need to do to avoid the possibility of nuisance power outages.

I will come to the details of why this will work shortly but, before we forget, it is important to look back to the third statement referred to above; namely we must be absolutely sure of the integrity of the Earth connection of the Tormach® machine, such that it is fundamentally safely 'grounded' and if any part of the machine's metal chassis were to inadvertently come into contact with mains voltage, a circuit breaker in our mains supply distribution board or in the machine itself would trip instantly due to the short circuit. It is probably important to add we would prefer the benefit of RCD protection for all our electrical equipment and it is a degradation in safety to bypass it, although in the special circumstances discussed, it is permissible. However, without RCD protection we must be confident the safety Earth system on the machine is secure and tested.

3.4.2  Step-down isolating transformer

There are two types of mains transformers which can step down the voltage, the first is called an isolating transformer and the second an autotransformer. It turns out that autotransformers are quite efficient and cheaper to make, so might appear the desirable option at first glance. But they offer no isolation whatsoever from primary side to secondary side.

We need the properties of an isolating transformer to solve the RCD problem. Remember the 32A ring main in the garage is protected by an RCD breaker in the distribution board for the house. If we connect this ring main, via a separate spur, to an isolating transformer which has a 230V primary (input) and 115V secondary (output) we can use that secondary voltage to directly power the Tormach® machine. But here's the thing; since the primary circuit is isolated from the secondary, what goes 'in' to the first wire of the primary always comes 'out' of the other wire. Nothing else can ever be true. Back at the distribution board, the RCD likes this very much indeed and thinks everything is working wonderfully.

However, on the secondary side, let's say the Tormach® is running at full power and, due to the nature of the VFD kicking up lots of harmonics, there is some leakage to Earth of the unwanted electrical interference thanks to the EMI filter doing a great job. Certainly, if we were to measure what comes out of one of the secondary wires and compare it to what goes back in to the other one, it is more than likely we would find small differences due to the fact the EMI filter has 'soaked away' some of the current down to Earth. In other words, not all the current coming from one of the secondary terminals makes it back to the other. But back on the primary side, there is no sign of any of this mismatch and the RCD device at the distribution board is blissfully unaware. We have effectively and successfully isolated the Tormach® power supply from the actions of the RCD breaker but crucially, everything else connected to the ring main (all our other things which are connected to 230V) are still fully protected.

In preparation for the delivery of the Tormach® machine, we need to purchase a suitable transformer and make the necessary connection to the garage ring main, but you will find there are different sorts of isolating transformers for sale: normal step-down transformers or Centre Tapped Earth (CTE) versions.

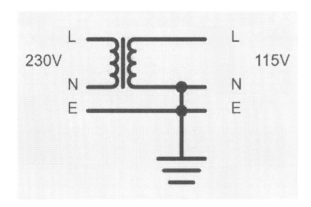

**Schematic diagram for a mains step-down isolating transformer**
Image prepared courtesy of SkyCAD Electrical CAD drawing software

If we could purchase a simple step-down transformer, that would be fine, but they are surprisingly difficult to source in the UK. This is because they sell far more of the version which has a centre tap in the secondary winding.

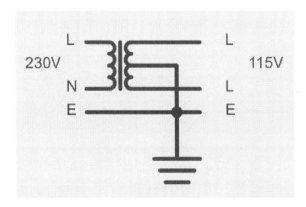

**Schematic for an isolating transformer with centre tapped secondary**
Image prepared courtesy of SkyCAD Electrical CAD drawing software

And the reason for this is the regulations on this side of the Atlantic dictate the use of 110V equipment on building and industrial sites for temporary equipment which must be connected by the so-called Centre Tapped Earth (CTE) method. This makes sense: you can imagine the hostile environment on a building site for anything plugged into the mains – cables getting snagged on sharp tools or being 'run over' by works' vehicles - and so they use step down transformers producing 110 volts and connect the centre point of the secondary voltage to Earth (Ground). This means effectively two live wires go to the equipment being used, but if a person were to touch one of them, it would only be 55 volts away from Earth, or Ground. It's a safe system, only superseded by battery-powered equipment, but even that needs mains to charge the batteries, of course, so these sorts of transformers will continue to be sold in large numbers.

You can see in the schematic diagram showing the centre-tapped transformer the secondary connections are labelled as 'Live' and 'Live' which sounds more dangerous than 'Live' and 'Neutral' but it is safer because each one can only be 55 volts different from Earth.

By now you are probably wondering why so much detail on transformers. Well, we might well have had to buy a transformer with a centre tapped secondary and then modify it. Importantly, we would need to slightly change the wiring inside the transformer box to achieve an electrical circuit like the schematic which has no centre tap, the Earth connection moved to one side of the secondary, and I shall subsequently be going into detail why this is so important.

I found an excellent manufacturer called Blakley Electrics Limited (based in Kent, UK) and, after much detailed discussion with one of their 'projects engineers', a gentleman called John Ackerman, I purchased one of their transformers. I mentioned how surprisingly difficult it can be to source a transformer with the secondary wired as 'NE' (earthed neutral and live secondary) rather than 'CTE' (centre-tapped earth) and I did end up having to modify the wiring in a 'CTE' transformer to turn it into the 'NE' sort. The part number I purchased was a Blakley S210264 and I spoke with John afterwards to thank him for his help and advice. Ever the gentleman, he extremely kindly offered to generate a new part number on Blakley's system which provided the exact transformer I would have needed, in the hope this may prove useful for my readers. The new part number is A7054566 which you will not find on their web page, instead you would need to phone him up or drop him an email. This transformer would power any of the 440 or 770 range of mills offered by Tormach®, the ones requiring 115V. It is a 2kVA continuously rated transformer.

Blakley Electrics Ltd., Part Number: A7054566
STLW/2/230-110NE/C1/IP44

Blakley Electrics enclosed transformer assembly supplied complete with the fitments detailed below:
Enclosure TL series, wall mounting, sheet steel, non-vented to IP44. The enclosure is phosphate pre-treated prior to electrostatically applied powder coating, shade Dark Admiralty Grey.
The wall mounting brackets can be repositioned on-site to make the enclosure floor fixing.
Rating 2000 VA single-phase, double-wound, continuously rated transformer, manufactured in accordance with BS EN 61558 Parts 1 and 2-4 but with an earthed secondary winding.
Primary Volts 230V
Primary tapping: +10V / 0V / 220V / 240V (These can be utilized to alter the secondary voltage if required)
Primary Protection to be provided on site up-stream, with an appropriate size and type of MCB or fuse.
Secondary Volts 110V NE
Distribution
Fitted with 1no SP 16A MCB Type 'C' to BS EN60898 protected under a clear hinged window (note DP MCB is shown in photo as part no. S210264 is shown)

**Blakley 2kVA wall mounted transformer similar to A7054566**
Copyright© photograph courtesy of Blakley Ltd

Key points are it is not an autotransformer, it is an isolating transformer, for the reasons we already outlined. And it is continuously rated at 2kVA (which you can think of as 2kW to all intents and purposes for the way in which the Tormach® will draw its power) and we already calculated our machine, the 770MX, should only draw about 1.5kW, so we are erring on the safe side.

Do please be aware there are some much cheaper transformers for sale which come in (usually bright yellow) fibreglass boxes with a moulded handle on top. You will see these on small scale building sites to power electric tools. These are a tempting option because they are inexpensive until you realise they are not continuously rated. They depend upon the use of the power tools on a building site being used intermittently, which of course they certainly would be. For this project, we are choosing a professional, wall-mounted, rugged, steel constructed box with a transformer inside which is continuously rated. In fact, the 2kVA transformers from Blakley Electrics Ltd. have an intermittent rating extending to 3300VA. It is important to do the job properly, and safely, so I strongly advise against purchasing one of those fibreglass boxed site transformers for this project.

The eagle-eyed reader will notice the transformer output is rated at 110 volts and we were hoping for 115 volts for the American

Tormach®, so what's gone wrong? The answer is absolutely nothing; it is well within allowable tolerance and, in any case, as we shall see when we connect it up, there are further connection options inside the box on the primary side we can utilise to fine-tune the final output voltage.

### 3.4.3  Connecting the transformer primary to the mains

Connecting the transformer primary side to the mains electricity supply is straightforward. If you wanted, you could use a flex with a 3-pin plug on the end and simply plug the whole thing into a 13A socket on the garage wall. There is nothing preventing you doing this – recall we already calculated the total draw from this socket would be about half that of an average kettle.

However, I decided to wire my transformer through an isolating switch to a short spur from the ring main. I bought one of those fancy rotary switches which have a slot in them for a safety padlock. In theory, if I wanted to, I could padlock the mains OFF thereby keeping things ultra-safe if I was doing serious maintenance on the machine. The isolator switch I purchased from RS Components was stock reference RS466-227.

**Mains-isolator switch from RS Components Ltd**
Copyright© photograph courtesy of RS Components Ltd

I decided to wire the transformer directly from my ring main, through this isolator switch, using 2.5mm² twin core plus earth proper mains cabling and connectors throughout. I secured the transformer above ground level, on a metal frame I fabricated which sits on the rafters beyond normal reach. The thinking behind this, and I have not seen any specific advice but it seemed a reasonable precaution, was the Tormach® has a large capacity coolant sump at floor level, containing some 12 US gallons of fluid. If there were ever to be a serious leak, I would much rather my electrics were not sitting in a large puddle, but rather high up out of harm's way.

When it comes to connecting to the primary side of the transformer you will notice it has multiple connection options. The manufacturer has added some extra terminals to the primary side to account for any variation to the primary input mains voltage and let you decide how you would like to fine-tune your secondary output voltage. I took a photograph of the inside connections within the steel box of the Blakley transformer to illustrate the options.

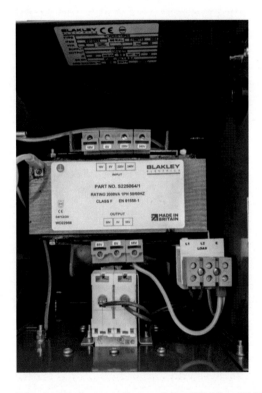

**The inside connections of the Blakley 2kVA transformer S210264**

The primary side is labelled INPUT and has 10V, 0V, 220V and 240V terminals. The transformer is designed to provide 110V on the OUTPUT so, for instance, if you had a mains supply which you had measured at 240V, you could connect this to the 0V and 240V terminals and still expect to get 110V on the secondary. But if you measured your mains was 230V, then you would need to use the 10V and 220V terminals (that 10V terminal is the 'other side' of the 0V terminal so adds to the total windings) effectively getting a '230V' winding and, again, should then give 110V on the secondary. Do you get the way it works?

Now, we wanted 115V for our American machine, not 110V. The UK mains electricity supply is supposed to be 230V these days to harmonise with the rest of Europe. But what they guarantee from the supply company is that baseline figure minus 6% or plus 10% as tolerance limits. This means it could be anywhere from 216 volts to 253 volts, quite a difference I am sure you will agree, but it's allowable so we won't worry too much about this. Let us assume the mains electricity in our setup is running at the target 230V. What we could do is connect this between the 0V terminal and the 220V terminal and, if we do that, we have applied a notional 230V where the transformer windings were expecting 220V, which is a 4.5% increase. Unsurprisingly, this will be reflected in the secondary output voltage which will now increase from its target value of 110V. As if by magic, 110V increased by 4.5% happens to be 115V.

### 3.4.4 Connecting the transformer secondary to the Tormach®

The transformer I purchased from Blakley Electrics (S210264) is intended for site-safety installations, thus utilising the centre tapped Earth connection. It comes from the factory pre-wired with the centre tap to Earth (CTE) which is exactly what we DON'T want. If you buy the special part number (A7054566) from Blakley Electrics which is already wired for 'NE' secondary, then your job is much simplified.

It is quite important to appreciate why you *must not* connect your Tormach® machine to a centre-tapped earth secondary of a transformer. And if the only transformer you can get hold of is one of these, you *must* convert the connections inside the box to the 'NE' configuration. Allow me to explain why this is so.

Built into the steel box of the 'CTE' transformers is a double pole circuit breaker, each pole connected to one of the two output wires from the secondary. If one pole trips, so must the other, due to a mechanical link. It is impossible for only one of them to trip. This is important for site safety when using 55V-0-55V centre tapped secondaries. Imagine a tradesperson operating a 110V electric drill. If there was a short circuit overload, then both wires will be disconnected simultaneously. That's important because if only one of the wires was disconnected, the electric drill would stop working (it couldn't work on 55V) and the operator of the tool could be forgiven for thinking it had been completely cut off, when this was not the case – it would still have half the voltage connected to it - a potentially dangerous situation.

Now, the Tormach® 770MX is a complex machine which has an electrical cabinet containing four separate circuit breakers. These are single pole circuit breakers because the Tormach® milling machine has been designed to be connected to a mains supply with a 'Live' and a 'Neutral' wire. All the single pole circuit breakers are connected in the 'Live' side of the supply, as you would expect. So, if a fault were to occur, any of them could potentially cut the 'Live' supply to their respective faulty circuit, rendering that part of the machine completely safe. Because they are not double pole, they cannot cut off the 'Neutral' wire, but in this case that doesn't matter because Neutral is connected to Earth (Ground) and is safe to touch.

So now consider if we inadvertently, and wrongly, connect the Tormach® mill to a CTE 55V-0-55V secondary, then the Neutral conductor is no longer being fed a safe 'earthed' connection. The metalwork of the machine would still be earthed (grounded) but the neutral connection into the machine would be 55 volts different to earth reference. And this would continue to be the case even if a circuit breaker inside the machine's electrical cabinet had tripped. And that is just not acceptable.

The new part number (A7054566), wired correctly as a 'NE' secondary output, has a single pole circuit breaker in the secondary, and this is specifically in the output marked Live. The output marked Neutral has no such circuit breaker as it doesn't need one.

Overall, there are three distinct mechanisms for electrical protection. The ring main is protected at the distribution board where the mains supply enters the property. The isolating transformer for the Tormach® is protected by its own circuit breaker in the secondary. The Tormach® is protected by its four main single pole circuit breakers. I trust by now you are forming the opinion we are well protected.

The Tormach® machines come with an American plug. I figured I wanted to keep this because it stops anyone inadvertently plugging the machine into anything other than an American socket. So, how do we get hold of an American socket, for use in the UK or Europe? Well, I am sure there are many ways but, as usual when I'm faced with a minor challenge such as this, I resort to RS Components Ltd and I found a wall-mounting double socket which is made by MK Electric. It is metal, with its own metal pattress box and looks tough and well-constructed. It is part number RS267-6558. The Tormach® 770MX plug fits snugly and firmly into this wall socket and that's quite important. Make sure you place this on the wall somewhere reasonably close to where you will be locating the machine because you don't want it vulnerable to somebody walking nearby tripping over it.

When you look closely at the shape of the holes in the wall socket for North American plugs you will notice the two rectangular slots for each plug are not the same size. This is because they are designated 'Live' and 'Neutral', with 'Neutral' being the larger of the rectangular pins on the corresponding plug.

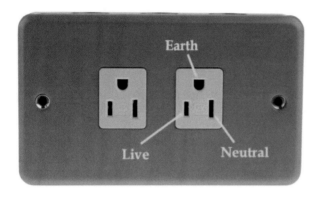

**North American mains socket made by MK Electric**
Copyright© photograph courtesy of RS Components Ltd

As you must now realise, it is important the transformer secondary output wire, which is joined to Earth, is the one we call Neutral. And it is vital we do not muddle up the Neutral and Live wires from the secondary. They must be wired correctly in the back of the wall socket. It is so important, this bit, I have made a little chart to help.

Americans do not have the same colour coding for single-phase electrical wiring as we do in the UK and Europe so this is going to be one of those rare occasions when the wrong colour coded wires will be in the back of a wall socket. All the more reason, then, to take extra care. The North American wall socket would normally expect *black* for LIVE (sometimes called HOT), *white* for NEUTRAL, and *green* for EARTH (always called GROUND).

| UK and EUROPE | | | North America |
|---|---|---|---|
| LIVE | brown | black | HOT |
| NEUTRAL | blue | white | NEUTRAL |
| EARTH | green/yellow | green | GROUND |

**Comparison of single-phase domestic wiring colours**

On this occasion, we will accept the use of UK/European wiring colours in this wall socket, but we *must* be certain the NEUTRAL truly is the one joined to Earth back at the transformer secondary.

If you get this the wrong way round, then you are creating a potentially dangerous situation in which your Tormach® 'Live' electrical bus within the machine is at Earth reference. If there was 'what should be' a 'Live' wire shorted to Ground due to a fault, the circuit breaker would not trip. Alternatively, if 'what should be' a 'Neutral' wire shorted to Ground, it would bypass the circuit breakers inside the machine and be a serious short circuit eventually causing the circuit breaker on the transformer box to trip. The circuit breakers within the Tormach® would be 'out of circuit' and offering no protection. This would be fundamentally dangerous.

Now we have the electrical supply wired up ready for the arrival of our North American machine, but there is one more consideration I have not mentioned up until now. For completeness, and just in case it applies to your setup, I shall discuss UK circuit breaker trip ratings.

### 3.4.5 Distribution board Mains Circuit Breaker: type B or type C?

Physics being what it is, at the instant you connect electricity to the primary side of a mains transformer, quite a considerable inrush current flows, albeit only for a few milliseconds. So much so, there is a risk of the main circuit breaker back at your distribution board tripping. We need to investigate this a little more to gauge the risk of nuisance trips occurring.

Normally in a domestic mains electricity distribution board a ring main would be protected by a 32A MCB or RCBO. An MCB is a miniature circuit breaker and detects long term overloads thermally, and short-term overloads electromagnetically. An RCBO is like an MCB but with the additional function of tripping due to Residual Current as we discussed before. So, an RCBO is a Residual Current Breaker with Overload Protection.

Thinking about the overload protection for a moment, if the rating for a circuit breaker is 32A then this is the maximum value of current the device will continuously permit before tripping will occur by the thermal tripping mechanism. But the electromagnetic trip mechanism is designed to allow short term surges, or spikes, much higher than the rating value. This is because many things in a household do cause a short-term surge so it makes sense to allow for a degree of very temporary overload. And there are different degrees to which a circuit breaker will permit such overloads, which are categorised as types B, C or D. Type B circuit breakers will allow between 3 and 5 times the rated value and are intended for domestic use. Type C circuit breakers will allow between 5 and 10 times the rated value before tripping and are intended more for industrial use or for when there is a specific issue which needs solving. Type D circuit breakers are heavy industrial devices not relevant for this discussion.

So, how much surge does a transformer cause at switch on? It depends on the state of the saturation of magnetic flux within the core at the precise moment of switch on, so in fact will be different every time you switch it on and, to some extent, the exact point in the waveform of the mains cycle the instant you switch it on. But worst-case scenario, a transformer could cause a very short-term spike in amps up to 10 times the rated current of the transformer.

Let's run through a very quick example using the 2kVA transformer I have been using in this chapter. Running at full rated power, this 2kVA transformer will have a primary input current of 8.7A (based on the fact the primary side is being fed 230V, and the product of amps and volts equals the power). Therefore, this transformer could cause a current spike of 87A (ten times).

Looking at my specific case, connected to my garage ring main which is protected by a Type B MCB at the distribution board rated at 32A, I can expect this MCB to trip if there is a short-term surge in current of 3 to 5 times that rated value, namely between 96A and 160A. I should therefore not expect to suffer nuisance trips at initial power up. (Note this is not powering up the Tormach® 770MX – rather this is the initial power up of the transformer alone). I can confirm I have had no nuisance trips, so the theory seems to work.

The reason I have mentioned it, however, is not everyone will be using a 32A ring main. Supposing, for example, your setup has you running your Tormach® machine from a 15A dedicated spur direct from the distribution board. This time, assuming a type B circuit breaker, you can expect it to trip for spikes in current between 45A and 75A, well below the 87A surge we think we might occasionally get from the transformer. So, in this case, we might suffer a nuisance tripping problem. A solution might be to change the dedicated spur to a 20A supply and use an MCB with a type C rating, which would increase the trip current trigger point. You might prefer to make the spur 32A with a type B circuit breaker, which we know works fine. You can see this is a discussion point, probably one to have with your electrician if you are not comfortable with this subject.

Finally, I wanted to address any concern you might harbour about plugging the Tormach® into a mains supply which is ultimately protected by a 32A circuit breaker. Isn't that a large circuit breaker for a machine we expect to be drawing only 6 or 7 amps? Well, I make the point again, we are quite happy to plug a table lamp into a wall socket connected to this ring main. The lamp undoubtedly takes less than an amp, and we rely on the 3A fuse inside the plug for local protection of the lamp fitting and trailing lead. In the case of the Tormach® machine, this has multiple circuit breakers within its electrical cabinet to protect various subsections within, so it is more than adequately safeguarded. Rather than simply plugging the transformer primary into a socket on the wall, it must surely be better to hard wire it, via a professional isolator switch. The good news is any failure of the transformer primary circuit shorting to earth would cause a residual current trip, as the 230V side is still very much protected in this way (remember it is only the 115V output we have isolated from residual current protection). Furthermore, any primary side failure such as the internal breakdown of its winding insulation would rely on the thermal overload protection characteristics of the distribution board circuit breaker, in my case the 32 amps limit on the ring main, for protection.

3.4.6  Power supply for a Tormach® 1100MX on 230 volts

It is worth mentioning the Tormach® 1100MX is slightly bigger and more powerful than the 440 and 770 versions. It runs from 230V American mains supply, so it is expecting 60Hz but will be fine on European 50Hz supplies. You may find it runs perfectly connected directly to your mains supply without all the hassle of transformers, but it is equally possible you could experience a nuisance trip issue due to the Residual Current Circuit Breaker at your main distribution board detecting nuisance leakage through the EMI filter on the Variable Frequency Drive, a problem we discussed earlier. The spindle motor on this machine is rated two horsepower.

You must not change the circuit breaker on your garage or workshop ring main to non-RCD type, just to get around this problem, because that would be potentially hazardous and doesn't meet regulatory requirements. In this case, you would be strongly advised, and better off, running a dedicated spur from your distribution board which has no RCD protection and leave everything else in your garage powered by the normal ring main. If that's not possible, or it is simply too much upheaval to install, another solution would be to use an isolating transformer, just as we have already discussed, this time not a step-down transformer, but one with a 1:1 ratio in windings, so 230V in and 230V out. Much as before, this would effectively isolate the residual current circuit protection on your ring main from the Tormach® 1100MX, whilst leaving it intact for everything else in the garage. In the case of the more powerful 1100MX you may need to consider buying a transformer which is a bit bigger and rated at 2.5kVA or even 3kVA, so things would get a tad more expensive. Also, the inrush current calculations will need re-visiting; bigger transformer equals bigger potential inrush.

Of course, it is worth re-iterating, if you have any doubt whatsoever about anything to do with the electrical supply, approach a professional qualified electrician. This section on 'electrics' will hopefully serve to raise your awareness of the potential problems installing one of these machines presents and will give an insight into the possible solutions but, unless you are totally qualified and competent to put these solutions into practice, please do get assistance and professional advice. Safety must be paramount every time.

On that final note, if you have made any fundamental changes to your domestic mains electrical system, then you should have the setup tested by a professional to get it 'signed off'. Specifically, if you have made changes at the distribution board, or to your ring main connections, then you will need to get the 'earth fault loop impedances' tested to ensure your safety earths are correctly connected.

It may well be the case a domestic electrician normally used to dealing with house wiring does not have prior experience of installing an electrical supply for a machine such as this, so it's quite possible the previous few paragraphs may well be useful discussion drivers.

## 3.5 Pneumatic air supply

The Tormach® 770MX needs a supply of pressurised air regardless of whether you are intending to fit the automatic tool changer or not. This is because the drawbar system on the spindle for this machine must have an air supply. BT30 tool holders are retained by spring pressure but are released by pneumatic pressure. There is no manual 'spanner' option. That leaves no choice; if you don't already own a compressor you are going to have to get one.

The manuals are patchy on the exact specification but, with a modicum of research, it seems the Tormach® is happiest when supplied with clean dry air at a pressure of 90psi and a delivery rate of 6CFM (cubic feet per minute). The machine does not need air very often, of course; only when you are changing a tool. So, to be honest, I think there is some scope for variation on these figures. Certainly, with an automatic tool changer your air demand will be greater. In fact, the Tormach® recommended 'California Air Tools 4610XC' air compressor only has a quoted air delivery of 2.2CFM so you can see it is all nicely vague.

Having said all this, it does need air often enough for it to be irritatingly noisy if your compressor is one of the older louder types. There are quite a few different manufacturers making more modern air compressors for the home user which are quiet, and the difference in noise level between the newer designs and the older ones is marked. It is truly worthwhile purchasing as quiet a one as you can.

There are a couple of other aspects worth considering. First off, if you are planning to install a Tormach® mill in the UK or Europe, I suggest you do not order the California Air Tools compressor from Tormach®. I am sure it's an awesome compressor but there are reasons why it is not sensible to import this from the USA. Maybe they will come up with a 230V version soon, who knows, but at the moment the only option is it works off 115V mains and we most definitely do not want to plug it into the same 115V supply we have just created for the Tormach® machine itself, because that is potentially introducing electrical interference on the very same cables. That's the last thing we want to do. Also, we carefully specified the 2kVA transformer rating for the Tormach® with some degree of leeway, but we certainly did not make any further allowance for the extra 750W of power the air compressor will demand. We don't want to burden the transformer any more than we need to. It's a bit of a no-brainer, we really must get ourselves a 230V 'quiet' compressor and run it from a mains supply as isolated from the Tormach®'s supply as can be.

When it comes to buying a simple air compressor for the home user there are a huge number of brands but, when you look closely, all the units are very similar, Chinese made, for the mass market, with different paint schemes to identify the importers' brands. I already owned a 'Sealey' branded 3HP V-twin compressor with a 50-litre reservoir tank. This is more than capable for running the Tormach® but, my word, it is noisy. So, I decided the extra few hundred pounds was worth the expense to source one of the newer quieter models. And that's when I came a bit unstuck.

It is true to say the American air compressor market for domestic use is so much more advanced than our European one. If you look online, you will find a lot more options and they are good at specifying pressures and delivery rates. Pop into any 'Harbor Freight Tools' store in the USA and there is a selection of compressors with a whole host of different and much higher output pressures, unseen in Europe. On most air compressors, certainly in the UK market, the air pressure switch, which governs when the pump motor switches on and off, appears to be of a standard design; a small black plastic box with a red knob on the top, which you pull up to start the compressor and push down to switch off again. Despite big warning labels to the contrary on the side of the tank you can, of course, make minor adjustment to the pressure at which the switch 'cuts in'. But the differential pressure, in other words the difference between the cut-in and the cut-out pressure, is normally fixed on these low-cost units, and is usually about 30psi. And that is fine if you start at 120psi and allow the reservoir pressure to deplete down to 90psi before starting the pump again to return to 120psi. We don't really want to go much lower than 90psi because that is what the Tormach® machine installation guide suggested we need. I have noticed on some slightly more expensive units you can have the option of a second adjuster within the pressure switch which adjusts the differential, so you could perhaps run the compressor between 95psi and 115psi, so a 20psi differential rather than the standard 30psi, for example. Again, there appear to be more of these sorts of options in the American market but are hard to find in the UK.

Anyway, I purchased a Hyundai compressor from an online store, which had a 50-litre tank, was 2HP, and was especially promoted as being quiet. Peculiarly, certainly from my initial viewpoint, it had two motors and two pumps working in parallel, so each motor was about 750W and the total load on my electrical supply was 1.5kW when it 'cut-in'. Why it didn't have one large motor with one large pump I have no idea, but maybe that's why it can operate so quietly. And to be fair, it is superbly quiet. I had also noticed on the store's web page this unit's specification read 'up to 100psi' pressure.

**Hyundai 2HP 'quiet' compressor with 50 litre tank**
Copyright© photograph courtesy of Hyundai

What I had not really thought through was, this machine would also have the standard pressure switch with no means of adjusting the differential and when they said it ran up to 100psi they meant it literally, so this machine pumped up to about 100psi and then depleted back down to 30psi lower than this 'book figure' before the motor would cut in again. Things were worse still, I measured cut out pressure at 98psi and cut in pressure of 68psi. I am sure it is well within reasonable tolerance for what was printed on its label, but this was starting to sound a little low for the Tormach®.

At this stage, my Tormach® had not been delivered. I was only at the preparation stage, don't forget, so I decided to leave it for now and see whether the Tormach® would notice the pressure not being quite up to scratch.

In a later chapter of the book when I get to discuss the assembly, particularly the part where the automatic tool changer gets fitted, the air supply becomes a crucial component in the whole operation, and I shall talk in more detail about some other technical difficulties I had to overcome and, also, some options you may well be interested in adding. For example, I added extra outlets on my compressor so I could have an 'air blast gun' hanging at the side of my machine, for blowing coolant and swarf from a freshly made part. You will need to do a little ad-hoc designing of your own air system to suit your own preferences, needs and desires. In all honesty, this comes together in your head much more clearly as and when your machine starts to take shape. So, let's leave this until we get to that section.

Another problem I had was radio frequency interference, I think, coming from the compressor motors, particularly when they cut out. Switching any large inductive load off always tends to cause electrical interference and this compressor, with its two motors, was no exception I found out. This interference can be 'in the air' as radio energy, and 'in the cables' as electrical energy, both of which can play awful havoc with the computer controlling the Tormach®. A good way to minimise the amount of interference generated by something like a compressor pump motor is to connect a snubber device across the mains at the point it is switched to the motor, in this case by the pressure sensing switch. The snubber device I used was an RS Components product listed as their RS813-430. This is a resistor and capacitor wired in series in one package and suitable for high voltages. When the motor is switched off, any electrical interference generated is quenched by this combination, preventing it travelling back down the mains cable, particularly at the precise moment the switch opens. These snubber components may be connected across the switch or across the motor itself, there is debate on which is more effective.

3.6  Spirit Level

Going back to the very beginning and looking at the machine footprint diagram in Fig. 1 by now you will have a fair idea of where you plan to place the Tormach® machine. It will sit with all its combined weight on four feet, each with individual height adjustment, and you can work out with some degree of accuracy, even before the machine arrives, where those feet will be placed. Consequently, you will have a good idea as to how level the concrete floor is at this location. The height adjusters on each foot are merely a large threaded assembly fixed rigidly to the underside at the four corners of the stand. There is no allowance for these feet to protrude from the machine's stand anything other than perpendicularly downwards so they will not take kindly to being asked to work on a slope. There is a hard rubber layer to allow for any minor irregularity but, the point I am making is, you will be best served if you offer your Tormach®, when it arrives, a flat and horizontal surface.

So, yes, obviously you will have a spirit level to hand to make sure your floor is flat and level. But the reason for mentioning spirit levels is more specific than that. When you get to install your machine, the very first item on the list of things to do is fix the four feet to the stand, then position the stand on its feet in the precise location you previously calculated, considering the requisite surrounding space for access. Next you must adjust the horizontal level of the stand as accurately as you can. And you will need a decent spirit level a couple of feet long to do it. This is because on the top of the stand, there are four attachment points onto which the actual milling machine will sit and be secured, and although the stand is a painted metal structure, these four top attachment points are accurately ground surfaces, paint free, which must sit exactly horizontally.

You will spend quite a long time – I spent about an hour - adjusting the four feet so that these four ground surfaces are exactly level. And you will use a spirit level fore and aft, left and right, all four options of those positions until you get it right. It is massively important to do so. And, of course, once it is done, you cannot move the stand at all. Because that will change the levelling process and you will need to start again. Trust me, concrete garage floors may look level at first glance, but they are not.

Don't forget the trick to using a spirit level accurately: always test for level between two points by trying the spirit level both ways round. This way, if there is any inaccuracy in the tool itself, you can eradicate the error. Think of it this way: imagine you owned an old spirit level which, when perfectly horizontal, the bubble was slightly off to one side. You wouldn't know this, so if you were trying to level something, you would adjust the surface to bring the bubble into the middle. But now, turn the spirit level round in the other direction. The bubble would move to one side, by twice the amount of the original error in your spirit level. So, to make this surface exactly level, using this particularly bad tool, you need to find the position whereby the bubble moves to one side one way round, and the other side the other way round, by the same amount. In a sense, you are averaging the error. Even if your spirit level is a good one, it is always a good habit to turn it the other way round and check the bubble remains in the middle in both directions.

Later, when you get to the exciting part of positioning the milling machine on top of the stand, you will have the first opportunity to measure the horizontality of the machine table. And because the table is ground flat with super precision, your 'builders' merchant' quality spirit level which you used to level the stand is no longer sufficiently accurate and you would be best to consider purchasing a machinist's level.

I made use of Amazon and purchased a '98-6' machinist's level made by Starrett, a high-quality engineering company based in the USA. The level is about six inches long, so of no use levelling the stand which I mentioned earlier because it would not reach between the four support points, but it is perfect for measuring how level the table is. And obviously you will do this as soon as you bolt the milling machine to the stand. Any fine-tuning of the machine can be 'dialled in' using the feet adjusters, but hopefully the original levelling of the stand was already very close. What is also important, is the machine table is level in the X and Y directions as you sweep across it – in other words, is the table twisted at all? And secondly, when you get the machine powered up and can electrically move the table in the X and Y directions, you can test for any twist as the table moves position. Tormach® explain this very well and provide an excellent video on YouTube on how to fine-tune this. They provide, in the delivery kit, some very accurate shims, which you may or may not need to use depending on your measurements, and these can be placed at any of the four support positions between the machine and the stand should you need to adjust for any twist. I have absolutely no intention of trying to explain the process as I would be unable to improve upon their excellent online presentation.

I am offering an image of the Starrett 98-6 machinist's level and you can see it has an adjuster nut on it. The instructions explain how you set up the level to be perfectly accurate to within one marker line on the bubble glass. And each marker line represents a change in angle corresponding to 0.005 inches per foot (five thou per foot) which is quite amazing when you think about it. And for all of us in Europe, that is less than half a millimetre for every metre.

**Starrett Company model 98-6 Machinist's Level**
Copyright© photograph courtesy of The L.S. Starrett Company

In preparation for the arrival of any milling machine you obviously do not need to buy one of these machinist's levels but I sincerely recommend you do because, in truth, it will be the only means of knowing just how accurately you have assembled it. And I think you really do need, and will want, to know.

3.7   Concrete floor load calculation

There is nothing written in the installation manual about floor loading but I thought it was worth mentioning as a discussion point. I happen to have a decent concrete floor in my garage but it might just be the case your intended workshop floor space is different construction or you have some concerns about whether the four feet might cause damage where they sit. After all, the entire machine, when it is finally assembled by you, will sit on just these four feet, nothing else.

I didn't have any real idea of how much stress I was about to place on my garage floor and so a few calculations to make sure I wasn't doing anything untoward seemed a good plan. I think you will be pleasantly surprised when you hear my results. Hopefully you can make use of this information if you are considering installing on a floor which is less structurally sound than a concrete base.

I saw somewhere in the manual it says 430kg all up weight. I forget where I found that now. That was for the Tormach® 770MX machine. You will need to know what the weight is for your machine if it's a different model, of course. Maybe there will be a couple of heavy vices bolted to the machine table, a steel part being machined, or a 4th axis unit secured in place. I don't even know if they included the mass of the automatic tool changer in their calculation of 'all up weight'. To allow for these additional masses and for the sake of this small exercise,

I shall round the total up to 500kg. Roughly speaking, then, we will assume we have half a metric tonne sitting on the garage floor.

Now, I have absolutely no idea how much my concrete floor can take without being damaged but it would certainly be interesting to calculate the actual pressure applied to the four points beneath where the feet sit. And then we could compare that to the pressure applied to the floor beneath the tyres of my car if I parked it in the garage. After all, the floor was designed for my car so, that would be a reasonable comparison.

Let's do the maths (for our American colleagues, we don't say 'do the math' like you do). A basic school equation suggests Pressure is Force divided by Area, where Force is just the Mass pressing against the floor on account of the acceleration due to Gravity. We could get into some heavy S.I. units' stuff here, but there is no need; let's do this in pounds per square inch of pressure (psi) because I think everyone has a 'feel' for pressure in those units, no matter which side of the Atlantic they reside. (I still can't get my head around checking my car tyre pressures in 'bar' - I much prefer to use psi.)

So how many pounds weight is 500kg? About 1100 lbs. And what is the total area this weight is sitting on? Well, each of the machine's feet is four inches in diameter, and there are four of them, so that makes a total area of 50 square inches.

The pressure applied to the floor is 1100 divided by 50 which equals 22psi, a remarkably low figure I trust you will agree. Putting this into perspective, the correct pressures for the tyres on my car are 35psi. That means for every square inch of tyre touching the concrete surface, there is 35lbs. pressing down. And yet, in our worst-case scenario, there is only 22lbs. of weight pressing down per square inch beneath the rubber feet of the Tormach® so I think we can safely say this will present no difficulties whatsoever. The floor is safe, but I hope you can see it was worth checking.

3.8 Coolant

Coolant, on the face of it, does not sound a very interesting discussion to be having. But it's a much bigger subject than you might first imagine, especially if you've had no previous personal experience of using it. I mean, I had used it at school on 'Colchester' lathes, and an old shaper they had, but the coolant was taken for granted back then, certainly by us students, in any case. Coolant is a beneficial addition to any machine as it contains rust inhibitors which protect the bare metal parts, the smooth flat surface of the table in particular. It can seem counter-intuitive, being 90% water, and you would be forgiven for thinking it might make things worse. Anyway, here we are preparing for the arrival of the Tormach® machine. We know we are going to need coolant at some stage and, like so many other things in this chapter, it pays dividends to sort these items out early. Once the machine arrives you will be so absorbed by the build process, and all the ancillary thoughts caused by having to solve occasional construction snags, minor purchases will become an unwanted distraction.

I researched coolant quite extensively online at first, and then found a few companies and asked some questions. I also saw some recommendations on YouTube videos. If, like me, you last saw coolant on a lathe at secondary school in the 1970s, you'll know it was a smelly and horrible, white-coloured suspension of water-soluble oil in water which looked like milk. Chemistry has improved since those days and the smell is much more controlled, and pleasant, and there are even some coolants which are virtually clear as water rather than milky to look at. And that can be quite useful for 'seeing' the part being machined more easily whilst being dowsed in coolant, which is particularly handy for those enthusiasts producing YouTube videos of course. The clear stuff turns out to be quite expensive though, I found out.

I found a supplier in the UK who, when I revealed I came with no prior knowledge, offered me a lot of advice, and suggested a coolant brand of theirs which was milky but not, shall we say, fully opaque and which would suit my needs and not be too expensive.

The supplier is Pennine Lubricants Limited, of Sheffield in the UK, and the oil they recommended is called Ultrasol X-777. It is a semi-synthetic oil meeting all the latest health and safety requirements and suitable for cutting both ferrous and non-ferrous metals, which sounded ideal for my needs. I highly recommend you purchase 5 litres of this in preparation for when your machine arrives.

And when the time comes to set up your coolant system in the machine you will need to add this product to water to produce a mixture which is about 9% concentration. The excellent advice I had from Pennine Lubricants also suggested it was best not to use only de-ionised water in the mix and that some tap water (preferably from a hard water area with calcium carbonate and other salts in it) was also a good idea. Apparently, this helps reduce foaming of the coolant.

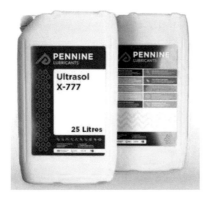

**Ultrasol X-777 semi-synthetic coolant from Pennine Lubricants**
Copyright© photograph courtesy of Pennine Lubricants Ltd

I purchased a 25-litre container of de-ionised water from Amazon and used tap water for the rest. I asked Tormach® what the capacity of the coolant tank was and they came back with 12 US gallons. It's important to realise a US gallon is slightly smaller than a British gallon, so be careful with any conversion calculations. As ever Google is your friend in these cases. I reckon my sump was a tiny bit less than 12 US gallons; so anyway, at the end of the day you must figure out how much water goes with how much coolant to get to about 9% coolant concentration.

You always add coolant oil to the water and never the other way round, otherwise there is a reaction and you get a lumpy mess. I knew this and added mine in the correct direction but still managed to get some inconsistencies. This only later showed up when the bits which had coalesced into lumpy solids blocked the metal gauze filter above the sump, where the swarf is collected, and my pump ran dry. I had to clean it all out, remove the solids, and try again. I believe I was not stirring it in well enough when I made my initial mix – it really is quite sensitive to this. Thankfully the problem never happened again, and I can report free-flowing milky coolant with no issues ever since that first attempt.

Of course, you are going to find it easy to set up an initial concentration simply by measuring the quantities of water and coolant you use in the mix, but the concentration will change with time and tends to get more concentrated. Here's why: There are two 'losses' from the mix; firstly, some of the oil mix is lost as it gets taken away with the swarf and sticks to it, never to return to the coolant sump. This is a very small loss but means you will need to add more of the coolant product in the future at some stage. Much greater volume of loss, however, is the loss of water from the mix, simply because water evaporates, whereas oil tends not to. And as it evaporates away into the air, that means the concentration will have increased. So, you need to top up with water every so often, because the total level is visibly reducing in the sump, and with coolant sometimes, to maintain the correct concentration. There is no way of knowing which you need to add, simply because you cannot guess the two loss mechanisms. Therefore, you will need a special tool to make accurate assessments of the concentration, and then make appropriate corrections to your mix.

## 3.9 Coolant Refractometer

Even though you probably think you don't need one of these, you do. Luckily, they are for sale on Amazon or you can order one from Tormach® when you order the machine. They all appear to look the same and be Chinese made. And they are not too expensive.

You may be familiar with these devices; they are used for measuring concentrations of liquids in other disciplines such as sugar content in fruit juices, anti-freeze concentration in car engines etc. They work by utilising the fact the refractive index of liquids changes with concentration of solution. You place a couple of drops of the liquid onto a glass screen, hold the eyepiece up to a bright light, or daylight, and read a number from the scale. The number you read is based on a special scale called the Brix scale and this is standardised.

Initially you test some pure water to make sure the number comes up as zero on the scale. There is an adjustment screw if this is not the case. So, in effect, what we are saying is we now have a 'Brix' scale, whatever that means we don't care, but Brix zero corresponds to pure de-ionised water, as a starting reference point.

In the case of machine coolant concentration, each manufacturer has done some tests to find out how their product changes the refractive index as the concentration varies and they publish a 'fiddle factor' to convert to the standard Brix scale.

So, in the case of Pennine Lubricants Ultrasol X-777 coolant, their datasheet, which is published on their own web site naturally, states that the Refractometer Conversion Factor is 1.4 which means whatever the number you read from your refractometer, you multiply it by 1.4 to get the actual value of concentration. In my case, I was aiming for a Brix reading of 6.5 so this would then give me a concentration of about 9%.

I purchased my refractometer from Amazon. You can purchase these with different scales.

**Refractometer and the Brix scale**
Copyright© photograph courtesy of Amazon.co.uk

You can appreciate in the CNC machine coolant world we are never going to be looking at particularly high concentrations so it is important you obtain a refractometer with a suitable scale. Some scales go up to 80 or 90 Brix, but when you are expecting a reading less than 10, then a full scale of 30 might be more appropriate.

3.10  Lubricating oil for the machine slideways

I had already had some interesting conversations with one of the experts at Pennine Lubricants Limited in Sheffield and I trusted their advice. It made sense to ask them what they recommended for the slideways lubricant for my Tormach® machine.

It doesn't matter whether you opted for the manual or the automatic oiler, the oil you put into its reservoir is the same. Slideways lubricants, you will not be surprised to learn, come in a variety of viscosities. There is something called the ISO Viscosity index and apparently the Tormach®, like many similar machines, needs a lubricant rated at ISO Viscosity 68.

Pennine Lubricants Limited supplied me with 5 litres of their product known as Slideways Lubricant TX68 which seems to fit the bill completely. It is reasonable not to expect any fluids to be included in the crates when the Tormach® is delivered, so you should expect to have to source all of them locally.

## 3.11  Pneumatics lubricant

The air pistons on the Tormach® which operate the powered drawbar system for the BT30 tool holders and the positional slide mechanism of the automatic tool changer (if you choose that option) need a lubricated pneumatic supply to keep them in tip top condition, so on the back of the machine Tormach® have installed a Filter Regulator Lubricator (FRL) which should accomplish this. It has a small reservoir which must be preloaded with special oil for the job. In the UK the workshop tools and supplies company, Sealey, call this 'Air Tools' oil and sell it in litre sized plastic bottles and I am fairly certain you can buy it through Amazon. The Tormach® is delivered with the FRL reservoir empty, again presumably for shipping reasons, so you will need to buy some of this in preparation.

## 3.12  Engine hoist

In the United States of America, Tormach® will come to your premises and install your machine for you if you wish. As far as I know, this service has been running for many years, although I believe there was a temporary pause in that process while the world adapted to the 2020/2021 pandemic. In the United Kingdom, I have checked with the managing director, Edd Clapham, at CNC Machine Tools Ltd., and they are now offering an installation service too.

However, if you are an 'engineering' type of person, which I am thinking you probably are if you have read this far, then all I can say is there is no better way to get to learn exactly how the Tormach® machine is put together than to do it yourself. It's an extremely rewarding process, (probably a huge understatement), and I genuinely recommend you give it a go.

There is just one single operation in the whole of the assembly when things become rather serious, and the consequences of getting it wrong quite expensive, and that is when you crane the milling machine itself onto the top of the stand and then secure them together.

Suspending the best part of 400kg of expensive metal more than two feet into the air with the help of a hydraulic hoist on castor wheels concentrates the mind, especially when you think of the price you paid for this machine and your mind starts to think of expressions like 'a chain is only as strong as its weakest link' and then you get a little bead of sweat forming on your forehead. Seriously though, it would be catastrophic to drop it on the floor and very dangerous if any part of anyone's body was beneath it. Some serious planning for this event is necessary in my opinion.

It depends on your particular circumstances but my installation began with a straightforward delivery into my garage which has a smooth concrete floor. I knew I was going to need an engine hoist for the assembly. Before the Tormach® arrives is the time to start planning this, because even the hoist will need assembly and it takes space to do even that. When Tormach® deliver your two crates on pallets, believe me, you will be running out of space to move.

There is an excellent YouTube video, which the Tormach® employees themselves have produced, which shows the entire lifting process using an engine hoist and their own 'Lifting Bar Kit' which I mentioned in the order options list in a previous chapter. You would do well to watch this a few times because it is worth committing some aspects of the intricacies of this part of the build to memory. You will glean a few things which will help in your preparation. When lifting the milling machine up above the height of the stand, which you have previously set up in the correct position and levelled accurately, there is not much space to push the legs of the engine hoist underneath the stand and you need to be extremely careful not to bump into it, for fear of upsetting the careful levelling you took time to accomplish. The engine hoist on metal castor wheels carrying the weight of the machine does not roll that easily, nor smoothly, has huge inertia, and is not easy to control. It needs thought, so start to plan this now, in your mind. That way, when the delivery truck arrives, you'll have some idea where you want the driver to put the two crates, and how you will bring the engine hoist 'into play'.

There are engine hoists with legs that splay apart at an angle and those which have parallel legs. It doesn't matter which you choose but bear in mind you will need to reach over the stand quite a bit as there is a limit to how far you can roll 'underneath' it if you have the splayed legs type, which seem more common, certainly in the UK. And if you have the sort with parallel legs, then you will need one in which the legs are far enough apart to reach either side of the stand because the crane will have to lower the milling machine down over it completely centrally. So that sort of hoist would need to approach the stand 'square on' from the front or rear. It sounds obvious but it needs thinking through.

**Foldable engine crane from SGS Engineering of Derby, UK**
Copyright© photograph courtesy of SGS Engineering Ltd

Some hoists come with an adjustable main lifting arm, such that the more arm length you adjust out, the less weight it is designed to carry, not only so it can stay in balance and not tip up, but also because of the leverage acting against the hydraulic piston. So, for example, I bought a very decent hoist which was rated at 2 tonnes lifting capacity, but when I set the lifting arm fully out, it then became a half tonne hoist, which was still sufficient for this task.

I chose a hoist from SGS Engineering (UK) Ltd., based in Derby, and they delivered it to my home address boxed and in parts which I had to assemble. This hoist was perfect for the Tormach® 770MX installation.

## 3.13 Pallet truck

Whilst you will most definitely need some means of hoisting the machine up onto the stand, you might be forgiven for thinking you can probably avoid the need for a pallet truck, so long as you get the delivery driver to leave the most important crate, the one with the milling machine in it, in a good position from which to extract the machine using the hoist. That's easier said than done, to be honest, and I strongly urge you to consider borrowing or purchasing a pallet truck. I found a very impressive supplier called Pallet Truck Warehouse of Milton Keynes who delivered me a brand-new standard size pallet truck. After the installation was finished, I sold it on eBay for almost what I had paid for it, so it really is worth the time and effort to plan and have this ready.

The item was called a Pallet Truck PT-04N, the N stands for nylon wheels, so it is hardy and the wheels are tough and do not deflect when weight is applied. The truck can move a 2500kg load so it is way stronger than needed, and you won't believe how inexpensive these are.

**PT-04N pallet truck from Pallet Truck Warehouse of Milton Keynes**
Copyright© photograph courtesy of Pallet Truck Warehouse

One of the most difficult things to do was cut away part of the pallet on which the milling machine was sitting, to gain sufficient access for the engine hoist to be able to extract it, and the pallet truck came into its own throughout this process. There is a lot of fiddling to be done, it is no exact science, and the pallet truck gives you the freedom you need to move the crate about and gain the access you will require.

I wholeheartedly recommend you consider getting one. And if you still don't believe me, then watch the Tormach® video again.

## 3.14 Reciprocating power saw

If you managed to find the specific video on YouTube which Tormach® produced showing the full assembly, then you will have seen the section near the beginning, when they need to literally remove part of the pallet the machine is sitting on. This is to gain the necessary access for the engine hoist to slide in close enough for it to be located above machine's centre of gravity so it can take the load. The pallet is from the factory in China and is made of tough stuff. Cutting through it is quite difficult and certainly not the breeze it looks on their video. You will need to have a reciprocating power saw so you can take bites from the pallet material aggressively whilst still with good control; after all, you cannot afford to damage the machine and there are cables and delicate parts all around. The pallet is made of ply type material and is full of nails at random so you need a tough saw which can take a little bit of abuse. I popped down to my nearest B&Q (Americans please read Home Depot for our B&Q) and bought a cheap Chinese power saw made under the brand Erbauer and it is fiercely powerful, comes with a few tough blades, and did the job.

It is called their ERS1100 model and the whole point of the reciprocating power saw design is the saw blades have an open unsupported end so you can literally power into a plank of plywood at an oblique angle with no leading pilot hole and it will take it. You need something as strong as that for this job.

**Erbauer ERS1100 reciprocating power saw from B&Q**
Copyright© photograph courtesy of B&Q

## 3.15 Allen keys in your toolkit

This might sound a strange topic to add to the list of preparations but unless you have a comprehensive set of Allen keys in your tool repertoire, please hear me out. Allen keys, otherwise known as hex keys, come in all sorts of formats these days, some are standard 'L shape' straight fit, some with 'ball ends' for using at an angle to get into confined areas, some with Tee handles, some come as 'bits' which fit into a magnetic screwdriver, and some are made to fit a socket wrench.

I am going to suggest you will need access to all these types during this build project, certainly in some of the smaller metric sizes, so make sure you have the tools ready for when the Tormach® arrives.

The other thing to mention is you will find a mixture of metric sized screws and bolts and imperial sizes too. That's not surprising being American designed, Chinese built, there is bound to be some overlap, so be ready for this too. There is nothing worse than using the wrong Allen key in a machine screw's hex-head and destroying it, and this is easy to do when you think something is metric but it is in inches. For the very large hex-head bolts securing the milling machine to the top of the base unit, Tormach® supply a suitable imperial Allen Key.

Things like the bolts which hold down the vice on the table, assuming you bought a Tormach® vice, are imperial, whereas the hex-head screws which are used throughout the machine's metal enclosure, holding it all together, are metric. For an American designed, Chinese manufactured machine it is all rather confused. But it doesn't really matter of course. Some of the hex-head screws are close against the inner wall of the enclosure prompting the need for a 'ball end' Allen key or a narrow screwdriver with an Allen key bit in the end. I highly recommend you have a screwdriver which has a good grip, is relatively narrow, and can magnetically hold 3mm and 4mm hex-bits, since it is better to approach these screws 'square-on' with a full hexagon shaped tool, rather than a ball-ended one, to avoid spoiling them.

Tormach® do provide a keyring with numerous small Allen keys attached; somewhere in the main delivery crate you will find a Tormach® 'tool bag' and I think that's where I found mine. Whilst these will get you started, please take my advice, and make sure you are suitably equipped with quality tools, especially for the 3mm and 4mm hex sizes. There are so many machine screws to tighten up, your fingers will be in shreds if you try to use some small and fiddly keys.

## 3.16  Metal snips, sometimes called tin snips

This may be obvious to some, but the crates which hold the Tormach® and all the ancillaries are bound up with steel bands and these are a breeze to cut with metal snips, but if you don't happen to own a pair of these, then metal bands are difficult things to deal with.

So, I would recommend you purchase a pair on Amazon. Not only for cutting the bands to gain access to the crates, but also useful for cutting the bands up afterwards into small lengths to make it easier to take to the waste tip.

While you are logged into your Amazon account, grab yourself a pair of safety specs and some protective gloves. Unpacking all these crates really is quite rugged, abrasive and splinter inducing. The steel bands can spring up quite suddenly causing a risk to eyes too. The Chinese make very robust crates, I shall say that for them.

## 3.17  American screwdriver

Are you still logged into your Amazon account? You may want to invest in an American screwdriver. I realise this might sound like the proverbial factory apprentice being sent to the stores on a hilariously non-sensical errand, but no I'm not joking. There are a couple of times when you need to undo and/or tighten up a crosshead screw and it doesn't feel quite right – the screwdriver is not snugly engaged with the screw. And you get that sinking feeling it's going to spoil the head of the screw. I'm sure you know the feeling. It is a fact of life there are several different standards for cross head screw designs around the world. In Europe and the UK, we tend to find things have gone the way of POZI-DRIV with different point sizes available. In North America, they are still rather fond of the PHILLIPS standard and, while they look similar at first glance, they certainly are not the same fit.

I am not criticising either one of them, it is just the way things are but certainly, if able, I recommend you have access to some different sized Phillips screwdrivers, just for those times your 'Pozi-Driv' screwdrivers don't feel quite right. In this Tormach® project, you are going to experience this feeling a few times and so forewarned is forearmed.

# Chapter 4    Delivery Day

4.1  Out of balance crates on pallets

Let me set the scene. The 18-tonne rigid heavy goods vehicle with a hydraulic tail lift arrives to deliver the Tormach® machine. The driver climbs up into the back of the lorry and switches on the battery powered pallet truck. The payload to be offloaded is two quite tall and weighty crates and, as he elevates the first one onto the pallet truck and guides it toward the rear of the lorry where the tail lift is, you notice something odd. The crate is wobbling precariously and appears not to be centrally balanced on his pallet truck forks. In fact, it is not even close to being balanced and the whole thing is tending to fall from the pallet truck to one side. With the tailgate now descending the driver catches the precarious nature of the combined pallet truck and unbalanced load with his full might just in time to avoid it tipping onto its side and falling from the tail lift completely. It is horrific to witness.

**The badly balanced crate**

You probably think it sounds like I am exaggerating this story for effect, but I genuinely am not. At first, I thought the driver had mishandled the picking up of the crate onto his pallet truck and then I looked at the pallet itself. The only way to explain is to show you exactly what I was seeing (see photograph).

Most UK pallets have three cross beams, with two gaps for the forks to go through. This pallet has two beams, not remotely central, and the gap between them is too narrow for the forks to fit inside. So, the driver is forced to choose a side and, whichever side he chooses, he ends up being way off centre and therefore completely off balance. It is shocking to watch when it is teetering on the edge of balance, only being saved by the swift brute force of the driver; especially so when you have just paid the final invoice prior to delivery. In the photograph, as the driver approaches the ramp, it looks quite level but he is holding it up with all his strength. The pallet continually wants to fall off to one side.

So, what went wrong? Once you open the crate and look inside you instantly realise the milling machine is very securely bolted to these two wooden beams; completely through them in fact, so they are in the position they are because they must line up with the four main fixing holes in the main casting. To lift this crate in a completely balanced way, you would need to lift from beneath the actual beams which, of course, the pallet truck forks can never do. A standard pallet truck forks are at a distance apart which will not go into the gap. This applied equally to the driver's professional pallet truck as to my one - the one I had just purchased - so the problem becomes a real issue moving this crate about the garage too. It is impossible to lift this crate evenly balanced and it will always be precarious.

On the plus side, the milling machine is extremely secure and it is impossible for it to move about in the crate, which is excellent for the journey across the oceans. But I think it is worth highlighting this to you simply because it really is a hazard when getting it off the back of a lorry with a tail lift. Both the crates have the same balance problem because the base casting is secured in a similar way with bolts through its crate's wooden beams.

The driver seemed surprised at how difficult it was to handle these crates so it probably wasn't a regular occurrence for him, which adds to the potential hazard, in my mind at least, because I could tell he was not expecting it at all.

## 4.2  Position crates carefully in the garage ready for dismantling

The larger of the two pallets is unwieldy with wide wooden boxes which take up so much space. Inside the boxes on top are all the enclosure panels, doors and windows which are incredibly well wrapped and secure. At the very bottom is the stand, a heavy metal structure, which will ultimately be the first thing you need to begin the assembly. In other words, you must be able to unpack all the other things on top, put them to one side out of the way carefully so they will not be scratched or damaged, then begin to work on this base part knowing you can move it past the other crate to where it will be sited for the final position of the machine.

Then you will need to be able to get your engine hoist to pick up the milling machine from the other crate and roll it to this location. You can begin to appreciate a little forethought and planning as to which crate goes where will place you in good stead. Your pallet truck itself takes up space and is not particularly manoeuvrable in a tight corner, so there is the potential for a conundrum.


4.3  Plan for rubbish disposal

You will be amazed at how much packing material there is which will need to go to the municipal rubbish tip. Much of it is heavy-duty plywood, with metal edging, sharp edges, and awkwardly sized to handle. It would be worth giving a little thought to how you will arrange this whole disposal process. This is not the sort of rubbish you can chuck in the boot of a family car; we are talking serious lumps of timber, with nails sticking out, and staples everywhere.

# Chapter 5     Assembly

## 5.1   Start of the build

I just checked the Tormach® 770MX Operator's Manual (which is their document 0720A covering everything from initial preparation through assembly to operation) and it is their Section 4, entitled Installation, in which they describe the assembly process in exacting detail. In this chapter, which I am calling 'Assembly', there is absolutely no point in my repeating what Tormach® do perfectly well in their manual. Instead, I will run through each of the major sections and reveal any top tips which I found from my experience of completing each stage. Although this amounts merely to bolting parts together, for the most part, things are never as simple as you may first think, so any snags along the way I shall call out. When you build your machine, I hope you will find it useful to keep track of where we are in this book, but obviously it is vital you will refer to Section 4 of your copy of the Tormach® operator's manual. All I can do is offer some extra pointers which I hope will make the process smoother.

    If you purchase a Tormach® machine, you will get a hard copy of its user-manual when it is delivered. But in case you are reading this book with a view to potentially purchasing a 770MX in the future, I would highly recommend downloading a pdf copy of their manual from the Tormach® website as soon as you are able: it is totally free of charge, and read in conjunction with this book, it will make a lot of sense and be extremely useful.

    Tormach® produce a lovely diagram which shows pictorially the assembly process in ten basic stages. (You can download it from their website and it's also part of the manual, section 4.1). It is a little out of date, in that it makes no allowance for those who purchase the newer PathPilot® Operator Console Assembly but there is a separate installation manual for that (Technical Document TD10715, again downloadable as a pdf for free). This diagram is a useful means of keeping a 'big-picture' situational awareness of where you are in the assembly process. If the PathPilot® Console option is on your list, it will come as no surprise to hear this is fitted last of all.

## 5.2 The stand

I already mentioned the vital technical drawing which Tormach® produce showing the floor space utilisation of the machine when it is fully built. So, you will already have marked out on your concrete floor exactly where the stand is going to be placed. Once you put it there, and have lifted the milling machine part onto it, it is not going to be easy to move. Almost impossible, to be honest. You will not be surprised to hear my first top tip, then, is check and double check all the possible interferences (walls and access ways) before you commit to the location and, when it is placed on its feet, spend sufficient time to level it accurately.

**Showing the stand with the coolant sump separated**

### 5.2.1 Front access

Make allowance for whichever controller you have purchased. The newer PathPilot® Operator Console, which is fabulous by the way, can swing out to some degree and, if you like the idea of using it at an angle slightly inclined towards the operator (who would normally stand in front of the machine doors), you might be surprised just how much space this uses. Because of that, I have chosen to fix my operator console facing outwards, perpendicular to the machine frontage, which works fine.

Secondly, I quickly found I needed an operator's tool bench/tool chest conveniently at my side when working at the front of the machine, especially for setting up tooling or work holding. I can confidently recommend the E72 range of storage trolleys from BAHCO, which are readily sourced from RS Components Ltd. These have a rubber lined top surface which sits 955mm from the ground, providing the perfect height platform for the operator to work from. You can choose from various options of drawer sizes to suit your storage needs. The units are all 510mm deep and about 780mm wide (including the handle) so, if you were to have it facing your left side as you work at the machine, you would need to allow approximately a metre of space in front of the machine so the trolley can still be steered out on its castors. It's all horses for courses, but you get my point. The BAHCO unit acts as an excellent side bench for the Tormach® operator and I promise you I am not being paid to say that.

**BAHCO storage unit acting as side bench for the Tormach®**

### 5.2.2 Left side access

You won't need any access to the left side at all since it is a fixed window and there is nothing you need to get to. At floor level, even with the machine right up against a wall, there is enough space to tuck away the air compressor, next to the automatic oiler which is bolted to the side.

However, when you assemble the metal enclosure you will, of course, need to tighten up the screws which hold it together, and those which hold the Perspex windows in place. These are all hex-head machine screws so you need to be able to wield an 'Allen' key with some dexterity if you have only left a tiny amount of space next to an immovable wall. So, another top tip: give yourself a four-inch gap minimum if you are up against a wall. You can probably see in the photograph, in my situation there is a garage door, not a wall, on the left side of the machine. I was able to open this door to fix the screws, so that gave me the option of reducing the gap to a minimum.

### 5.2.3 Right side access

You would be forgiven for thinking the right side of the machine needs no access. After all, the window is also fixed this side. But there is a cupboard built into the right side of the stand and the PathPilot® Console does extend further to the right than the sides of the enclosure, so I suggest taking heed of the recommendation from the technical drawing. Also of obvious importance is the electrical cabinet which has plug in connectors which protrude from the right side and this is where the main electrical ON/OFF switch is, so you do need some space here.

### 5.2.4 Behind the machine

Do ensure you allow enough space, as per the drawing, behind the mill, since you will be popping round the back more than you might think. This is where the pneumatic air supply comes into the machine, where the main access door to the electrical cabinet opens, and where the removeable panel behind the automatic tool changer is located. Also, if you intend using a 4th axis unit, such as Tormach®'s microARC4, this has a cable which will need to be fed through to the rear of the machine and plugged in.

I would suggest you will be well placed if you can allow full walk-in access behind the machine.

### 5.2.5 Clearance above

During normal operation, when the milling head moves up and down, the height of the machine varies of course. The very tallest point is the black plastic flexible 'chain' which feeds all the electrical cables, slideways lubricating pipes, pneumatic lines, and coolant hose through to the head. Tormach® refer to this as the 'energy chain', and clearly you need to ensure your ceiling height does not interfere with this. Equally, you should plan to have adequate space for your engine hoist hook, together with the Tormach® 'lifting bar kit', to lift the machine to at least an inch or so higher than the stand. This is so you can move the whole assembly - swinging on the hook as it most certainly will be - without knocking your carefully placed stand out of alignment or adjustment. I shall come back to what this height is, imminently.

### 5.2.6 Fixing the feet to the machine stand

This is all covered in the manual but what they neglect to mention is the welding seams on this large base assembly may protrude somewhat proud of what was intended to be a flush surface in the design, meaning that when you screw a foot to the base, it does not sit squarely to the surface. Each of the four legs is bolted into a gusset of steel at each corner and the way in which they have welded each gusset in place means the welding intrudes into where the leg should sit flush. If this is ignored, with the weight of the machine on each foot, a sideways force is imparted on the threaded securing bolt which will make it very hard to adjust. It is obviously important to ensure each foot sits square to the floor. I made sure to clean away the excess weld material with an abrasive flapper-wheel fitted to my hand-held power grinder so the feet were flush to the base and truly perpendicular to the floor.

**A good attempt at initial levelling of the stand**

When you have fixed the four feet and positioned the stand in the final resting place, be sure to spend a good deal of time levelling it with a spirit level across the machined faces on which the milling machine will eventually sit. Don't forget to turn your spirit level around and check it in 'both directions' to account for any error – you really want to get this as horizontal as is possible. When you have done that, you then need to avoid knocking it ever again until the mill is lowered down onto it and secured. Unfortunately, this is easier said than done.

5.3   The milling machine itself

This is probably the most stressful part of the whole operation since, for an appreciable time, the 'expensive bit' is hanging from a chain. You need to be utterly careful and methodical. You don't want to damage it, of course that goes without saying, but also you don't want to have any accidents; the weight of this machine could cause serious harm if it fell, so take good care and never place yourself in potential harm's way, beneath it when suspended.

**Lifting the mill onto the stand concentrates the mind**

In the photo, you can see the mill is still clear of the stand, by an inch or so. Note how beneficial it is for the mill to be perfectly horizontal whilst hanging free. This aids the lowering down into place without imparting any sideways forces which might knock your stand position and spoil your level feet adjustments. The Tormach® lifting bar kit obviously helps in this, but you still need to make some fine adjustments to get it 'just right', by trial and error when you lift it from the pallet.

During my installation, shown in the photograph, the absolute top of the engine hoist came no higher than about 2.3 metres (91 inches) so that will give you a good idea of the vertical space you need for the installation. In normal operation, once the 'energy chain' is re-fitted and the machine is running, you will not require any more height than this. The Machine Footprint Diagram supplied by Tormach® confirms this requirement for 2.3 metres height although, strangely, they seem to think that's 88 inches.

5.3.1  Releasing the mill from the pallet

I already mentioned about getting a power saw to help you cut away part of the pallet to gain sufficient access with the engine hoist to lift the machine from the pallet. It is messy, difficult, and hard to express in words how you go about this so I wholeheartedly suggest you view the videos which Tormach® have produced for their YouTube channel.

Having fixed the Lifting Bar Kit to the top of the machine and gently taken up the slack you will 'test' to see how close you are to getting the Centre of Gravity beneath the hoist so that ultimately you will be able to lift the machine exactly horizontally. If you do this correctly, it then makes lowering the machine down gently onto the stand in the right place much easier.

Firstly though, the machine is still bolted to the remains of the pallet you haven't sawn away. The idea is you get in close enough to lift the assembly, remove the bolts and remaining pallet parts, then lower the machine carefully onto something flat like a piece of hardboard avoiding damage to the machined faces on the underneath.

With the machine now unencumbered by any pallet material, you must refine your 'lifting point' to get it exactly horizontal. One thing I found which is not mentioned in the manual, the lifting tackle does interfere with the 'energy chain' to a small degree. To avoid any possibility of mechanical damage to it, I removed the securing screws so the chain assembly fell away loosely. It doesn't detach, of course, being still held by the cables and pipes within it.

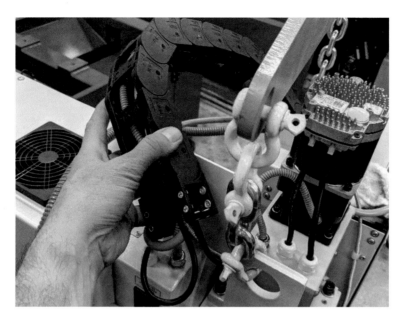

**Disconnect the 'energy chain' to make way for the lifting chains**

5.3.2 Lowering the mill onto the stand

It goes without saying, the coolant sump needs to be well out of the way so the engine hoist can roll beneath the stand sufficiently far enough that its 'cargo' can be lowered precisely squarely and horizontally into place. You may choose to practice this operation with an empty hoist – I certainly found that a useful exercise. You need to drive the hoist at an angle if your hoist has splayed legs and, again, the Tormach® video helps a good deal to visualise this. It makes a difference I am sure if the legs are of the parallel type, causing very different access problems.

It is most satisfying when you have the four large securing bolts in place, the weight is relieved from the hoist, the chain dangles loosely, and all is safe and secure.

What I find confusing, Tormach® suggest this moment is a suitable time to use the supplied shims to level the bed (or table) of the machine. Clearly, if the stand was horizontal, the bed will be predominantly the same. The shims are for fine-tuning of any 'twist' which could be apparent in the frame. You will be able to measure to see if there is any twist when you can power the machine up and move the table in the X and Y directions. The problem is later when you add the enclosure and other accessories such as the automatic tool changer, these additional heavy weights might affect any twist in the frame. So, I am guessing now is not necessarily the best moment to worry too much about this, despite access to the four main mounting bolts, and access to be able to leverage the mill up sufficiently to add a shim, never going to be better. The only reason to be considering adding shims at this stage of the assembly would be if you could identify and measure some discrepancies between the four main machined 'pad heights' of the stand. It's a bit of a puzzle, for sure. Luckily, I found my machine to be extraordinarily square and I never did find the need to add any shims.

When securing the mill to the stand, it's important to avoid uneven tightness on the four bolts, as this in itself could cause some flex of the machine. I don't know of any prescribed torque values, but sensibly tight, and all the same would be a good starting point.

Fine-tuning of any twist in the machine is always possible at the very end with very accurate cutting tests, but I never needed to get into this.

5.4   PathPilot® controller and first 'power up'

If you have chosen the PathPilot® Console, you cannot attach this to the machine for quite a while yet. It's one of the last things you do, because it screws into the enclosure after that is screwed together.

However, you most certainly do need to power up the machine and make sure all is well, before proceeding with the build. The tests are documented in the manual and need no further discussion here – you are just making sure the 3 axes function and the spindle motor turns, but also it is the first time your controller firmware will be booted up so you need to go through the initialisation process to make sure PathPilot® is up and running successfully. In my case, I had the Wi-Fi dongle attached and immediately I told the machine my Wi-Fi router password, it was chomping at the bit for an update from the Tormach® server. All very impressive. And a nice cosy feeling. Time to put the kettle on. The stressful part is over.

My recommendation for this part of the process is you will need to power up the machine a few times over the course of the assembly so you would be best placed if you take time to plan for this. The problem is you don't really want the PathPilot® touchscreen lying on the concrete floor while you are stepping over it with unwieldy metal enclosure pieces in your hands. It is a recipe for disaster and you can ill afford to damage the screen. The controller box must be relatively close to the machine so the cables reach the connections. It gets a bit untidy. And the touchscreen is clearly vulnerable whilst horizontal.

I set up a 'Black and Decker Workmate' next to the machine and placed the touchscreen lying flat with a piece of plywood cut from the original packing box carefully shaped to be a screen protector which I religiously made sure was covering the screen unless I needed to power it up. In this way, I was able to be extremely careful of this very important part of my machine despite its obvious vulnerability during the rest of the build process. You wouldn't even want to risk putting a screwdriver down on that screen, let alone inadvertently knock it while carrying another part of the metal enclosure. Take the time to protect this vital component.

## 5.5 Automatic oiler

The installation section calls for the setting up of the automatic oiler. I guess even if you opted for the manual oiler, you would still need to bolt it to the machine and connect the pipework, but there won't be any setting up as such.

It should all be very simple. The oiler is powered from one of the sockets on the rear of the machine, meaning it gets a 115V supply whenever you power up the whole machine. And it has a built-in timer controller with a digital display. The recommendation in the manual is to set the oiler to pump for 12 seconds every 480 minutes. That's 12 seconds of oil feed into the arteries of the machine to lubricate the slideways once every 8 hours. Not much, but apparently enough. The oiler will also pump for 12 seconds whenever the mains is first switched on. I found the oil pipes at the very top of the machine didn't reveal an oil flow at first (you can pull back the spring protectors and view the oil through the translucent pipe). So initially you may well need to disconnect one of the pipes near the end of the flow, somewhere near the head's slideways (the Z axis) so the oil must come over the top of the energy bridge, and then run the pump continuously until oil is fully primed into the circuit. Then re-connect the pipe. That worked fine for me.

I should add I also had initial confusion causing me to email and eventually phone up the helpline at Tormach® in the USA. The oiler supplied with my machine appeared to be broken and whenever mains power was supplied it would run continuously – it wouldn't even allow me to programme the timer. After many weeks of frustration, it turned out to be a known problem, about which Tormach® were fully aware. Except nobody had bothered to put any warnings into the installation manual, nor tell any of the helpline staff about it.

Slightly frustrating, it turns out, inside the oiler's electronic box, beneath the lid, which is securely screwed on, there is a tiny 'jumper' on the circuit board which overrides everything and causes the timer to be locked out. Removal of this jumper returns it to normal operation. I guess this would be a useful feature if you were trying to remotely control the oiling function from afar. The original equipment manufacturer of the oiler decided to supply the unit with this jumper in situ and it is just a little disappointing they chose not to put a sticker on the outside advising this. Lesson learned.

## 5.6 Pneumatics

The OEM Filter Regulator Lubricator (FRL) supplied by Tormach® was a cheap model and I could not get it to work. It just refused to prime itself, even on max adjustment with continuous air flow, and would not sustain any drip rate of oil into the system. I spoke with someone at Tormach® who admitted they had received other reports of similar issues. I decided to upgrade my set up.

The pneumatics within the machine runs two systems, which are very much inter-related: the air-powered drawbar and the automatic tool changer (ATC). When you install the ATC unit, you make slight modifications to the pneumatic pipes for the drawbar simply because the automatic function of the ATC 'takes over' the operation of the drawbar too. That make sense, of course. It has all been very well designed and carefully thought through and is all set up using 1/4-inch nylon pipe with push-fit connectors. I was not about to meddle with any of this but I did wish to purchase a better quality FRL for the machine and I chose the Draper Expert unit which is model number 24326.

**Draper Filter Regulator Lubricator #24326**
Copyright© photograph courtesy of Draper Tools Ltd.

Naturally, being a UK supplied unit, the inlet and outlet threads are British Standard Pipe (BSP) threads, and this is where you suddenly realise it could all get very complicated, very quickly.

A little bit of research using Google is quite revealing; there are two distinct camps fighting it out over pipe threads. Over in North America they are using what they term the National Pipe Thread (NPT). In the UK we use the British Standard Pipe thread (BSP). These threads, size for size, are remarkably similar, having the same or very similar pitch, but the angle of the thread form is 60 degrees for NPT and 55 degrees for BSP. They almost do fit together, but you must never mix them because the small difference will cause leaks. It is more a problem on account of their similarity, that they sometimes do get intermixed.

What is truly interesting is where the rest of Europe sits on this. Surprisingly, they choose to use the BSP thread too, which is even more remarkable since their metric thread system for normal screws is based on a 60-degree thread form. Nobody would expect the Europeans to use the expression 'British Standard' so they renamed the threads 'G' type. Thus, 1/4-inch BSP is equivalent to a G 1/4 thread. They are one and the same thing. But they are definitely not the same as 1/4-inch NPT. The Atlantic Ocean divides us completely on pipe thread systems and that is not going to change anytime soon.

And then you consider pneumatic piping, which uses hard but flexible nylon tubing and push-fit connectors. In North America, as you would expect, the typical sizes are imperial, such as 1/4-inch pipe and 3/8-inch pipe. In this case, they mean the overall diameter, or O.D., of the pipe. In the UK and Europe, the nylon tube and the associated push-fit connectors come in metric sizes, such as 6mm and 8mm etc. Again, these refer to the outside diameter.

It is important to note it is easy to purchase adapters for all combinations you can think of. In other words, I could get a screw-in adapter which is 1/4-inch BSP male thread (which would fit my Draper regulator nicely), the other end of which is a 1/4-inch push-fit nylon pipe connector. In this way I could properly connect the outlet of my UK sourced FRL to the 'quarter inch' pipe that feeds into my American Tormach® machine.

However, the inlet of my FRL will need something different. I decided to supply the air from my compressor through a manifold (which incidentally gave me other options but more on that later). From this manifold, the supply to the FRL inlet was a 6mm nylon pipe. So, the other adapter I needed was a 1/4-inch BSP to 6mm push-fit connector.

You can see the FRL is effectively the 'demarcation point' between NPT and BSP in my setup. The FRL is the 'Atlantic Ocean', the divide, between the North American mill and the European compressor. It is good practice to clarify in your mind where this switchover is, to avoid any mismatching of pipes and connectors.

One thing I must highlight to you is 6mm nylon pipe and 1/4-inch nylon pipe look the same. They are extremely difficult to tell apart (except by measuring accurately with a vernier) but you must not put 6mm pipe into a 1/4-inch push-fit connector. It will possibly come loose and fly out, and it will certainly not have a good seal. You won't be able to push a 1/4-inch pipe into the 6mm push-fit connector. It won't fit. Please don't be tempted to try. A good idea is to always purchase imperial nylon pipe in, say, black colour, and always metric pipe in, say, blue colour. I chose that way round because Tormach® have already fitted their machine with black pipe. Having a system like this will save confusion.

Don't forget to put special 'air tools' oil into the Filter Regulator Lubricator and prime it fully. Then adjust it so you get just a drip each time the ATC does a tool change. The whole point of spending some extra cash on a decent Filter Regulator Lubricator is to keep the powered drawbar piston, and the automatic tool changer piston in tip-top condition. And that is going to keep your machine running reliably.

## 5.7 Flood coolant system

Supplied with the machine is a coolant pump from Taiwan called a TC-8180. The first '8' indicates it is a 1/8 horsepower pump. The subsequent '180' reflects the fact 180mm (seven inches) of the pump projects beneath its mounting flange, to submerge into the fluid.

The outlet from the pump is threaded 1/2-inch NPT. Tormach® supply an adapter to convert to 1/2-inch push-fit nylon tubing. Exactly like the pneumatics system but larger bore.

It appears to be the case Tormach® originally intended for the 770MX machine to be fitted with a 1/4 horsepower pump. You can tell this because the coolant pump mounting tray has a hole which exactly fits the flange and mounting screws for such a pump. The tray on my machine had an adapter plate already screwed into place reducing the size of the hole to fit the smaller pump supplied. The larger pump which would fit without the need for the adapter plate is called the TC-4180.

I ran my machine for a short while with the originally supplied pump but later decided it would be worth upgrading to the higher-powered pump. I am pleased I did this as I am getting appreciably better coolant flow. One of the problems associated with using an American machine on a transformer is you can't change the frequency of the supply. The coolant pump has a simple single-phase induction motor, so its speed is directly related to the supply frequency. This side of the Atlantic, running on 50 Hertz (or cycles), the pump motor is rotating at 84% the speed it would have been if it had a 60Hz supply, so this directly affects the pump flow rate. Going for the pump upgrade negates this effect completely. And the extra electrical load is accounted for in the design of the circuit on the controller printed circuit board inside the Tormach® electrical cabinet. It is relay operated and has a 3A fuse, so well within design limits.

**TC-4180 coolant pump upgrade fitted**

I am not that keen on the way Tormach® have designed the coolant pipe system as it emerges from the pump. The problem is the pump is bolted securely to the sump 'drawer' which, in turn, is mounted on small castor wheels. It can be pulled out from the front of the machine to inspect, empty the swarf (chips), replenish the coolant, and gain access to the pump and the oil skimmer. To accommodate this movement forward and aft, they have used a 'recoil' pipe which springs back into a tight helical-coil shape as you push the sump drawer back into position. This connects directly to a 1/2-inch push-fit connector on the pump. And to be honest, when you pull the drawer fully out, there is a lot of tension pulling on that connection. If that were to fail one day, without anyone realising, the next time the coolant pump was switched on, things would get extremely messy very quickly. I decided to modify my set-up to put my mind at rest.

I purchased a length of EPDM (ethylene propylene diene monomer) rubber hose which had an internal diameter of 10mm. I then got hold of a brass adapter which was 1/2-inch NPT thread to a barbed taper suitable for a 3/8-inch ID hose, and I secured one end of the hose to the pump this way, with a 'Terry clip.' I used this hose to allow for the 'in and out' movement of the sump, by allowing some slack, and at the rear of the machine I converted back into the 1/2-inch nylon pipe. The conversion was done using a brass tapered barb to 1/2-inch female NPT threaded adapter, coupled to a 1/2-inch NPT male thread combined 1/2-inch push-fit connector. It all sounds very confusing but it alleviated my concern. And the EPDM hose is extremely flexible even when cold, which is what I needed. And it works great. But you must convert back to the nylon pipe because that then goes into the machine, through the 'energy chain' to the nozzles and it wouldn't pay to start changing that design. It is just where the sump can move, I wasn't happy with the flexibility afforded by their original design.

There are some online sources suggesting EPDM rubber is incompatible with hydrocarbons, especially petrol, but I made some enquiries and I'm advised that water soluble oils do not appear to be an issue for it. Time will tell on this. If I find appreciable degradation over the medium term, I shall convert to 10mm bore reinforced PVC hose, which is slightly less flexible but 'defiantly' robust to most chemicals. PVC pipe, though, tends to harden even more with age and use. For the time being, I am sticking with the EPDM rubber hose.

Whilst still on the subject of coolant, there is one more tiny problem I needed to solve. Inside the coolant sump drawer assembly, next to the coolant pump, is a space reserved for the tramp oil skimmer. This was one of the options we mentioned in an earlier chapter. The brand of skimmer supplied by Tormach® is an American one called 'Skimpy.' The unit is well made and should perform very reliably but I found a minor issue with mine which was resolved through discussion with Tormach® and Skimpy themselves, extremely amicably it must be added.

I had been sent a 115V version of the oil skimmer with an American plug on the end of its flex. I ran it from the same transformer supply as the Tormach® machine. The problem is these units use a shaded-pole motor with quite a low starting torque and running it at 50Hz in the UK was having a detrimental effect and the motor was very often stalled. The specialist at Skimpy agreed that in future Tormach® should consider recommending to all UK and European buyers to take the 230V version of their oil skimmer as this was rated for operation at both 50Hz and 60Hz, whereas the 115V version was only rated for 60Hz operation. Tormach® arranged for me to be sent a replacement motor and this has worked flawlessly ever since.

## 5.8 Automatic tool changer (ATC)

The automatic tool changer should be renamed the game changer. It is an awesome piece of machinery and I mean that sincerely. Tradition would suggest, my being an Englishman, I ought to have reservations about using the word 'awesome' but in this case it is entirely the correct choice.

I know it can always be argued you don't really need the automatic tool changer. That is true, of course. If you are operating at hobbyist level or doing complex prototyping work, you are going to be operating the machine very much 'hands-on' and can change tools manually quite easily. But if you can stretch to it, I implore you to buy this option. Once set up, it works reliably, without fuss, and makes a comforting industrial noise as the pneumatic valves sequence each tool change. More importantly, once you have built up a useful library of tools, you will find you leave commonly used ones in the carousel and can call upon them with ease.

**The ATC bolted into position**

The installation and set-up are carefully and accurately described in the user manual. The fine adjustments are crucial so that the tool holder lines up with the drawbar system, ensuring reliable tool changes and minimising wear and tear.

**Setting up the ATC fine adjustments for reliable tool changes**

I had some issues worthy of note for anyone intending to install this unit, which I'll describe.

5.8.1  They changed the design of the mounting bracket

Firstly, the ATC is a heavy item and is fixed to the machine by initially securing a large metal bracket to the side of the mill. The drawing labelled Figure 4-50 on page 57 of the Operator's Manual (version 0720A) does not accurately depict the bracket supplied with my machine, the 770MX. I am guessing there has been some design re-development with which the manuals department have not caught up. It's hardly important, you might think, but it did catch me out a tiny bit.

**Tormach® re-designed ATC mounting bracket**

The bracket is held on by a complicated system of threaded studs, four stand-offs, and nuts. How much these studs stick out from the side of the mill is down to you and, looking at the drawing, you wouldn't think it that important. However, on my bracket which is a different design to that shown in the manual, the two uppermost studs protrude ABOVE the platform upon which the ATC sits. And if the studs protrude too much, they interfere with the side of the ATC electrical cabinet, such that i) it doesn't fit and ii) it scratches the paint on the cabinet quite badly. It takes two people to bolt the ATC into place because it's so heavy; one must hold it in place while the other secures the bolts. When you are trying to figure out between you why it won't fit and having to hold this weight up in the air while doing so, inevitably the paintwork gets scratched. It is a bit disappointing. If only someone had mentioned the re-design and the fact the upper two studs need to be almost flush with the nuts, so they do not protrude at all, then this would all have been a lot less frustrating.

### 5.8.2 Electrical interference on the ATC USB cable

The second issue I experienced with the ATC was electrical interference causing the main controller computer to glitch and freeze. If I unplugged the USB cable from the ATC, this problem seemed to go away. It is a long USB cable to reach across the width of the entire machine and, I am guessing, is prone to picking up electromagnetic disturbances. I knew I had a potential source of those from my air compressor switching off each pressure cycle; I had already fitted an 'R-C' snubber component to the compressor so I had to think of another solution.

    I was discussing it with a colleague and he suggested buying some split ferrite sleeves, the type which you clip round a cable – they have a snap to lock mechanism making them permanent once fitted. They look just the same as the ones you see already moulded onto some USB cables, a sort of 'bubble of plastic' normally about an inch long. I found a source of these on eBay, a pack of varying sizes, and fitted one of these at either end of my ATC USB cable. There was absolutely no science to any of this. I have no idea of any inductance figures or anything like that. All I can say is it worked! The problem disappeared.

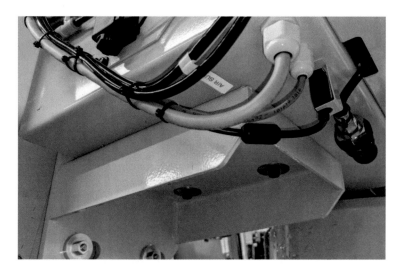

**Showing the ferrite sleeve added to the USB lead from the ATC**

You will recall Tormach® make a big case about running the machine from an independent supply and I knew I was still running my compressor from the same ring main ultimately, so this was hardly ideal, I admit. It is a tricky thing to solve if you only have one mains supply in your garage, as I do. Anyway, maybe this is the solution. It seems to have worked well for me. Thank you to my colleague for the suggestion and thank you eBay.

5.8.3  Air blast nozzle problem

I experienced a rather shocking moment when I first tried to use the brand-new ATC with a tool holder inside it. I had gone through all the set-up routine which basically gets you to use the software to finely adjust the position of the 'Z' height of a tool holder when it is locked in the spindle to match the ATC height (which is fixed) and then set up the rotary aspect of the carousel so it is able to receive the tool holder accurately into one of the receptacles. The BT30 spindle also has dogs which must align exactly for the tool change operation, so it must rotate automatically to achieve this. It is very clever.

Everything had gone well and the ATC could load and unload a tool holder adeptly. The problem came when I had left a tool holder in the carousel and it rotated to the next position; there was an almighty 'crack' and a piece of plastic pinged off from somewhere inside the ATC and fell into the bowels of the machine. I retrieved it from the gauze filter of the sump. It was a sort of plastic nozzle.

I partially dismantled the ATC to gain access to find out what had just happened. The best access was from below and by removing three large bolts the carousel can be completely removed. You learn a lot more about how things work when you take them apart, of course, and as soon as I removed the carousel, I could see what had happened. But at this stage I didn't know why. There are two air blast nozzles which are intended to blow compressed air at the tapered shank of the BT30 tool holder to clear any debris before storing a tool, or before loading a tool into the spindle. I know there has been some discussion on forums about the fact they use the same air supply from the ATC for this so, in effect, blowing 'lubricated air' rather than clean dry air at the tapered shank. Personally, I think that's a good thing, but the argument from some quarters is you are squirting oil into the fray which will end up as tramp oil in the coolant sump when it gets washed away. I can live with that. We are talking minute amounts of oily mist.

For some odd reason, one of my air nozzles had been knocked off its fitting when the ATC carousel revolved while it had a BT30 tool holder within its grasp. My investigation and subsequent discussions with an engineer at Tormach® in the USA, Norman Kowalczyk, found there had been an error in the factory assembly of my automatic tool changer. Tormach® produce three variants of the automatic tool changer; on a Tormach® 440 machine, the ATC has an eight-position tool carousel, on 770 machines it has ten positions, and on 1100 machines it has twelve positions. On that basis alone they are different, but there has also been some re-design of the air blast system, and the 440-variant had 'air blast' simply from two cut ends of pipes held in place by a bracket, whereas on the larger machines, the pipes connected to proper moulded nozzles.

To simplify the 770 machines apparently, there had been a production change to remove the moulded nozzles and just having open ended pipes instead. Somewhere along the way, the factory in Mexico got muddled up and fitted nozzles to my ATC but on the wrong bracket so that the tool holder crashed into them as soon as it was first used. I suspect there might have been a batch of ATC units built with this error, so I tell the story here in case you find a similar issue and are finding it difficult to interpret.

I decided to make my own re-design, to keep the proper nozzles and make them work. The alternative was to accept the re-designed bracket from Tormach®, dispense with the proper nozzles, and just have pipes pointing roughly in the direction intended.

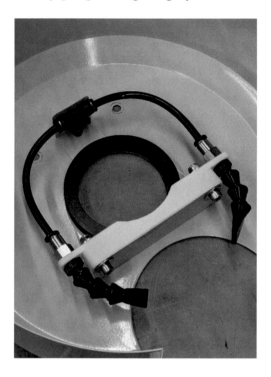

**White 3D-printed bracket re-positions the two air nozzles correctly**

I rather like the idea of proper air nozzles blasting the air at the tool holder tapered shank and the suggestion an open-ended pipe could be as well-aimed and directed is nonsense. I find this sort of re-design is often an accountant-led operation rather than an engineer's suggestion. I wanted my machine to have accurately focussed air blast, so I decided on a 'set-back' polycarbonate bracket to mount my two nozzles so they no longer collided with the tool holders. I am very happy with the outcome, which works well. By keeping the 'LOC-LINE' (American trademark) pipe and nozzle on each side, it remains fully adjustable. It is just that everything is now set back by about 15mm.

I think it goes to show how fast Tormach® is evolving, trying to create great products and, sometimes, there will be production issues. I have no problem with this. I accept they have costs to control too. They are so open about their products and are always keen to listen, take you seriously, and try their utmost to fix things. We are all engineers –

otherwise we wouldn't be buying one of these machines – so sometimes we need to step up to the plate and sort things ourselves. That's how I choose to see it.

What I can guarantee you, the fundamental Tormach® machine is an amazing piece of equipment. A little bit of fine-tuning isn't going to hurt anyone and you end up with a machine that is even more refined, and one you comprehend in ever more detail, all of which becomes immensely satisfying. I have included the details of this minor issue, even though I may well be the only person to ever experience it, because I guess their intention is still to produce future tool changers with the simplified 'pipe end' nozzles, and you may wish to improve upon that.

## 5.9  Building the enclosure

It must be said the quality of all the steel pressings that make up the enclosure is phenomenal and the assembly is an absolute dream. One person can do the whole thing alone, and it is extremely satisfying work. The directions in the manual must be followed exactly in the correct order.

The sturdiest two pieces form what Tormach® call the 'Chip Pans' and these bolt to the machine stand to form a horizontal lip upon which all the side pieces can sit. Also, they internally create a sloping chute down which all the coolant and swarf can slide to the 'chip catcher' which is a large rectangular box with a metal gauze filter sitting suitably above and within the coolant sump.

**Half the metal enclosure is assembled**

There is a lot of fluid splashing around this whole assembly and, unsurprisingly, unless you can be sure every single hole, gap, crevice, join and screw hole has been sealed, you will end up with puddles on your floor in no time at all. Ask me how I know this!

I don't think Tormach® make it clear enough in this part of the installation instructions just how important it is to be fastidious with the sealing. They supply some black sticky stuff on a roll which they call 'butyl tape' and suggest you place it between all the joints but I didn't find there was sufficient for anything other than just the chip pans, and certainly not enough for higher up.

You will not have any appreciation for how good at finding leaks coolant is until you switch the flood coolant pump on AND aim the nozzles directly at a large cutting tool, like a facing tool, which is spinning fast. That's when you will get serious spray all around the inside of the enclosure and that is what you need to do to see if you have missed any more leaks.

But I want to give you a top tip. Tormach® supply some rather strange translucent neoprene washers which are presumably for using with the pan head screws which secure the chip pans to the machine stand. As soon as you tighten the screws down with a hex driver, they 'squidge' out from under the head of the screw and fail disastrously. They are unfit for purpose. There is a better solution, and you need to know what that is.

There are some beautifully made stainless steel washers available online which have a neoprene type seal physically moulded to one side. The rubbery material is suitably hard so when tightened down under a pan head screw, with the seal facing downwards of course, it stays in place, does not 'squidge out', and creates a perfect seal with the steel enclosure beneath it. This is helped by the fact the washers are slightly concave to the seal side, this tending to 'capture' the rubber material within its grasp as it is tightened. These are exactly what Tormach® should be including in the kit. But they don't, yet.

If you Google *EPDM or Neoprene A2 STAINLESS STEEL sealing roofing washers*, you will see the sorts of washers I mean. I found some on eBay. The larger screws which secure the enclosure walls to the chip pan are 6mm diameter and the smaller ones which join the enclosure walls to each other are 5mm diameter. You can certainly purchase these sealing washers for both sizes.

**Stainless steel concave sealing washers from eBay**

All the side wall enclosure panels are secured to the 'chip pan' base by pan head hex drive screws and all require some sort of sealing washer (which are not supplied). The problem is the holes in the enclosure are elongated slots to allow for tolerances, which is understandable, but if the slots are visible at the side of the screw head after installation, then the coolant will find a way through. So, you must have a sealing washer wide enough to cover the entire slot.

The largest slots – and Tormach® are aware of this – are on the brackets which form part of the sliding door mechanism. You must be rigorous and look around for all holes and slots and make sure you have sealed them up. As far as the vertical joins between neighbouring side panels are concerned, I found the design to be excellent and I didn't need sealant between them, the folds in the steel designed in the correct way to direct the coolant spray back inside and down the chute. I have read on forums other owners have chosen to use builders' silicone sealant between their enclosure wall panels. The problem with that is it can be quite a strong adhesive and difficult to dismantle at a later stage. Better, in my opinion, to find other 'washer' type solutions which cover the slots completely. You must make your own choices here. On a couple of the slot positions inside my enclosure which were causing difficulties, I printed some specially shaped plastic washers to ensure the slots were fully covered.

**The enclosure and internal lighting finished**

It is probably true to say you will still find some tiny coolant leaks after a few hours of machining. At this stage don't worry too much about the finer details, if you have made sure all the visible slots are covered. It will be a case of reacting to further tell-tale drips on the floor as and when they appear. It's just the way it is. Once you have gone through the process fully, after a few hours of operation, then you can expect the problem to remain reliably solved.

## 5.10 PathPilot® console assembly

Finally, all the different parts of the enclosure, Perspex windows, LED floodlights, doors and 'roll-in' sump are all assembled, and the machine is looking amazing. Except for one thing: the PathPilot® controller box with the touchscreen is still lying on the 'Black and Decker workmate' with a piece of scrap plywood for a protector. And the cables are all over the place.

Luckily there is a separate section in the operator's manual to tell us how to construct this next piece of the jigsaw. The first thing it tells you to do is remove the right window. Having just fixed this into position, this is somewhat frustrating and makes you wish you'd read a few pages ahead.

I'm certain it was not intentional. The 'new' PathPilot® Console is a recent upgrade and they simply haven't fully incorporated it into the assembly processes of the machine. What this also means is this is the first time you are going to have to drill holes in the shiny new enclosure you just constructed. And it feels kind of bad if I'm honest. Up until now, all the screws have fitted accurately into pre-drilled and pre-threaded holes. But this time it is a matter of drilling clearance holes for the bolts and securing them with nuts and washers. You will need to take the right-side window out to do this. The Operator's Manual is not wrong. If you have a helper, you might think you do not need to remove this window as they can surely reach through the front by sliding the door open to gain access. But I'm afraid it's not that simple.

Take your time, check, and check again before drilling holes in the shiny painted metal, and take extra care to heed the warning about keeping the right door closed while you do this, so you don't drill through your door. And you need the right window removed so you can gain suitable access to place washers and nuts on the back of the screws. If you tried to gain access by opening the door, you would cover the screws you needed to fit nuts to.

The new console design works well and is over-engineered if anything, made of solid steel, and is quite heavy. I am sure it was a minor oversight but I could not find any screws, washers, or nuts in my delivery pack. In the scheme of things, this was no issue at all, in fact it gave me the opportunity to order some nice 'NYLOK' self-locking nuts as suitable fasteners. There is not much room on the inside between the sliding door mechanism and these retaining nuts, so a thin washer is all you can really afford to fit, so self-locking nuts makes complete sense. And, of course, the bolts must not be too long.

**The PathPilot® Console main support pole attached**

According to the Tormach® Operator's Manual, we have reached the end of the assembly stage.

In front of you should be a gleaming machine, looking magnificent, shiny, and clean inside, and out. We are soon to switch on the controls and be doing messy stuff like turning on the flood coolant and cutting metal. Take a photo of your new creation, put the kettle on, sit down and enjoy the view. A moment of self-congratulation is the order of the day. You will never see the machine this clean again.

# Chapter 6  PathPilot®

Whilst assembling the machine, it has already been necessary to power the machine up a few times along the way, allowing the more inquisitive amongst us to explore a little of the PathPilot® Console touchscreen. The best way to 'get to grips' with any operating system is to use it, not read about it. That's why this chapter is very short and the next chapter is much more fun. The plan is to start cutting metal as soon as (safely) possible and learn on the job. More practice, less theory.

PathPilot® is an incredibly well-written interface between human operator and CNC machine. I have not learned any other systems, yet I feel I can still say this with some conviction. Why? Because of the speed and relative ease with which I picked it up.

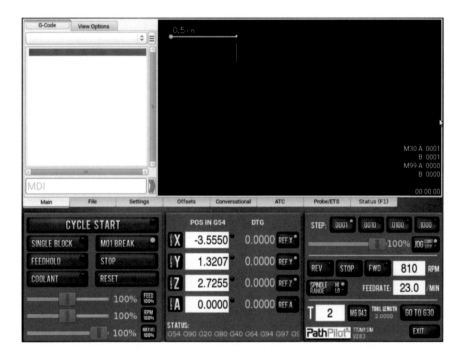

**PathPilot® Console touchscreen with the Main tab highlighted**
Copyright© screenshot courtesy of Tormach® Inc.

Since spending time getting to know PathPilot® intimately through using my Tormach® 770MX I have found it to be the epitome of user-friendliness. The much over-used phrase 'it does what it says on the tin' is perfectly apt.

I don't need, nor wish, to go into any debate about the alternative systems out there. If you purchase a Tormach® CNC machine, you are going to have to learn to talk in 'PathPilot®', so that's the only system I plan to discuss. We will learn as we go, but I just wanted to 'chat' through a selection of the more interesting aspects.

6.1   General synopsis

On the PathPilot® Console option, the screen supplied is a touch screen. If you choose to use your own computer without this feature, then the mouse-controlled cursor does everything your fingers can do on the screen. I like the touch screen controls immensely but you do need to be cautious of transferring oily fingers from inside the machine to the screen.

There appears to be no haptic feedback, not even a beep, when pressing buttons on the screen, which I feel would be a nice feature. Maybe someone knows how to make it beep for button presses; please let me know if you find out.

Just beneath the touch screen on the grey metal console, in addition to the 'EMERGENCY STOP' switch, and main electrical 'RESET' switch, there are two buttons marked 'CYCLE START' and 'FEED HOLD'. There are also three variable control knobs for manually over-riding the machine's speeds. All these controls feature on the touchscreen too, but it is useful they are replicated as real mechanical switches and knobs to reduce the wear and tear on the screen, with mucky fingers.

The screen is divided into two halves, the lower half never changing its format, the upper half adapting according to which of the eight tabs has been selected. You can see the tabs along the centre horizontal of the screen.

You could write a book about PathPilot® and working through each of those eight tabs would be a great starting point. But there is no need. It all makes sense as you begin to experiment. Some of the tabs you will already have visited in setting up the extra add-ons, such as the probes. Much of the set-up work is 'once only', so if you need to re-visit these aspects, it will always be safer to resort to the Operator's Manual for guidance. Those regularly used areas of the software quickly become second nature. Here follows a few salient points I have found from my initial experiences.

## 6.2 Status line

There is a string of G-code data at the lower centre of the screen, which relate to the overall STATUS of the machine. All these codes are discussed later. Just remember this is where the machine tells you what MODE it is in.

## 6.3 Zeroing the axes manually

If you press any of the ZERO buttons in X, Y, Z, or A, you are immediately overriding the origin position of the associated axis in the Workspace the machine is currently in. In the screenshot, just above the 'zero' buttons, you can see the machine is in G54 Workspace. Be extra careful not to do this inadvertently; it will cause you grief. Unless you are using the machine manually, you are much more likely to be referencing your work offsets with the passive probe automatically.

## 6.4 Current tool indicator box

At the lower right side, you can see the number of the tool currently in the machine. In this example, it is indicating tool number 2 is in the spindle. When you come to insert the Passive Probe into the spindle, this is where you must enter the value '99' manually. When you remove the probe tool – you can't ever store it in the ATC since it has a trailing lead attached to it – you must delete the value '99'. If you forget, the machine will soon tell you.

## 6.5 Go to G30 button

There is a 'GO TO G30' button on the screen just near the Tool number box. G30 is what I have termed a 'safe space' which you program yourself, a useful feature to prevent tool crashes, and something which I shall fully explain later. Getting to grips with G30 is particularly important if you are doing automatic tool changes, and I shall write about how you can and why you should use this feature to your advantage. (To program your G30 location, there is a button in the Tool Offsets tab.)

## 6.6 Status tab

When something goes wrong, the Status tab turns yellow, so you can't miss it. The Status tab button is labelled 'Status (F1)' because if you touch the F1 key on your keyboard, the screen will reveal that tab instantly, and show you what the problem is.

## 6.7 Store Current Tool command button

A very useful button marked 'STORE CURRENT TOOL' is on the ATC tab. I find this the best way to load tools into my tool changer. Simply, I manually load the tool into the spindle, type its number into the Tool number box, bottom right, and then on the ATC tab, press 'STORE CURRENT TOOL'. In the Tormach® user manual there are other ways of loading tools into the machine, but I found them confusing. This method couldn't be simpler. In fact, after the initial setup of the automatic tool changer, the 'STORE CURRENT TOOL' button is the ONLY button I ever use on the ATC tab.

## 6.8 Avoiding a crash

Naturally, there is plenty more to learn, but it's probably easier and more fun to find out more about PathPilot® just by using it. Nonetheless, it is evidently important to make every effort possible to avoid crashing the machine. By this I mean driving the spindle, tool holder, or cutting tool, into the work, the fixturing, or even the table itself, with such energy we would cause damage. It is surprisingly easily done, especially when you are new to this. I have crashed my Tormach®. It is a horrible feeling I can tell you. You will be amazed how quickly you can thump the emergency stop switch when you really feel the urgent need. Get into a mindset of always double checking what you are about to do, trying your utmost to avoid a crash. I have another top tip for you to take onboard before we continue.

**Showing the manual controls just beneath the touchscreen**

Just beneath the screen, the three rotary knobs are marked 'FEED RATE', 'RPM', and 'MAX VELOCITY'. The 'max velocity' knob, when you are starting out, is your friend. When you press the cycle start switch the spindle moves down towards the work at such a rate that if you had miscalculated your work offset in the Z axis the tool could hit the work or the vice well before you had time to reach for the Emergency STOP. To help you, initially, wind down the 'velocity' knob to very slow, maybe even as low as 5%. This is the velocity at which the machine will then perform all its motion. Not just feed rates, but rapid motion too. Wind it down very low and then, keeping your fingers on it, press the cycle start button to tentatively begin the new operation. If you find, in this slow-motion action, the spindle is heading dangerously towards a crash, simply wind the knob down to its lowest position and the machine will STOP. It is not an emergency stop, but a stop, nonetheless. If you wind it back up again, the machine will continue. It is a truly powerful control which allows you the opportunity to be doubly sure things are not going to crash. And it gives you time to make the decision. I cannot overemphasize how useful this knob is.

And, by the way, if you are left wondering how to get out of the sticky situation when you have stopped it from crashing by winding the velocity knob all the way anti-clockwise, slowing it right down to a complete stop, here's what to do next. Press STOP. Then press RESET (the one on the screen, not the power button below the screen). Full manual 'jog control' is regained, so you can drive the spindle up out of the way safely, under manual control, either with the jog handle or with the keyboard shortcuts.

Whilst on the subject of resets, I thought I might explain the difference between the large reset button next to the emergency stop button, and the small reset button on the screen, lower left. Think of the large reset switch with the blue light as the 'power on' button, and the emergency stop switch as the 'power off' button, both set up as a normal 'no volts release' (NVR) circuit. Power to the motors in the machine cannot be established until the reset switch is pressed, and you can hear the contactor as it closes. The reset button on the screen, on the other hand, is a reset of the current program code in the computer memory. If you stop a program mid-way through a run, because of an error or something, by pressing reset on the screen, the cursor will return to the beginning of the current program and wait there until the next time 'cycle start' is pressed.

## Chapter 7  First cuts

### 7.1  So where to start?

Eager as you will be to cut some metal – or 'make some chips' as the Americans say - you simply must cover a few essentials before you can bring a cutting tool to bear against metal the very first time. And this involves setting up a few datums. (It's funny it doesn't sound so cool for us Brits to say, 'make some swarf'.)

In the simplest terms, let's consider a piece of metal securely clamped in vice jaws waiting to be faced off. In the spindle, we have a milling cutter in a tool holder. Two things need to happen before we can bring the two together. We need to tell the machine controller where exactly in 3D space that piece of metal is. And we must tell the machine controller where exactly in 3D space the tip of the cutting tool is. Only then can the control system calculate the path for the two to meet safely.

So where indeed do we start? It's an interesting question and I plan to introduce you to this section with a slightly different approach to some other books I have seen.

Let me set the scene from my perspective; I had assembled the machine and knew in detail its mechanical workings, but I suddenly realised I knew almost nothing about getting the machine to cut by itself. To be frank, it was a little daunting. I literally didn't know what to do next. From the old-fashioned engineering knowledge inside my head, I knew how to operate a manual milling machine. I knew what milling 'was', for want of a better way to say it. From my more recent experience of the 3D printing world, I could also understand the layout of 3D space within a machine and I could do a reasonable job of modelling a part in the 'design space' of Fusion 360 software.

But there was a troubling gap in my knowledge. And if I am quite truthful, I just played around with the machine learning by trial and error and making mistakes along the way, with my hand hovering close to the emergency stop button whenever there was any doubt in my mind about what I was doing. Which was almost all the time, at the beginning. I do not think this was a particularly efficient way to learn how it all worked. I respond well to being taught things by someone on a one-to-one basis but in this case, I didn't have access to anything like that. I had the Tormach® operator's manual, a decent resource, and I could seek out useful YouTube videos to teach me more, but neither of these replace someone telling you the fundamentals face to face, enabling you to ask questions back.

If you'll permit me, I plan to bombard you with some basic facts you absolutely need to know to operate this machine. From there you will be well placed, to move forward at a more efficient pace, to learn the detailed stuff that is going to be relevant to your specific needs.

The other thing is, I am going to choose one method for each task we need to accomplish and that's all I shall talk about. Let me explain what I mean by that: for example, the first thing we will do is discuss offsets. We will start with work offsets, then move onto tool offsets. Most publications will go into detail about all the various ways you can measure offsets. I am not going to do that. I don't see the point. I figure you don't really care about the finer details of how or why different methods, of doing the same thing, work. Surely, you'd like to hear about the newest way things are done, not a history of all the ways. You would prefer to get up and running and have your machine making parts. It's a bit like saying,

*"you've bought a new phone and I am going to give you a quick lesson on the basics of sending someone a message using the 'WhatsApp' app, but for the sake of completeness, first I am going to talk about the basics of the older system we called 'SMS' texting. And I am going to remind you this is still a viable way to send someone a message, and then later we will discuss other methods of sending someone a message, such as email, before finally moving onto the finer points of WhatsApp."*

When I said I didn't want a book about the history of CNC machining, this is what I was getting at. I only want to suggest to you the cleverest, easiest, most modern way of doing things. In many ways, I totally admit I am leaving gaps in your knowledge, but it's deliberate. My intention is to get this machine working for you. Let's talk about Work Offsets to start us off.

7.2  Work offsets

Work is the stuff we are going to machine, metal, plastic or whatever. We are going to start off with a piece of STOCK and we are going to end up with a PART. That's an important distinction, and when we drew the PART in the Design space of Fusion 360, we chose an origin.

The origin is where the three axes meet, and the PART 'exists' somewhere in Design space relative to that origin. Once the PART is designed, we move to the Manufacturing space in Fusion 360, where we will be 'performing' Computer Aided Manufacturing (CAM). And this ultimately generates the G-code for our machine to make the PART out of the STOCK. It is obvious but I am going to say it anyway: the STOCK is larger than the PART.

It is crucial for you to appreciate that when you move to the Manufacturing space, you can put the 'manufacturing' origin anywhere you like. It does not have to be where the designer put it, back in Design space.

Many parts you wish to make will have two sides. And you'll machine one side first, then turn it over (flip the part seems to be a common expression) and machine the other side. Why must we flip it? Simply because we only have a 3-axis machine and we must hold the work somehow while it is being machined and it's impossible to machine all six sides without having to let go. Obvious, again, but worth stating, nevertheless.

Let's call this a 'two-setup' part, and we will try to generate two separate pieces of G-code to perform these operations. We will call these 'Setup-1' and 'Setup-2'. When we flip the part in the vice ready to do the second operation, we will use a different origin reference in the manufacturing space. Each operation needs its own reference in space, its own Work Offset.

So, let us consider Setup-1 first; place a piece of metal in the vice, tighten it up, and there it is, our STOCK, sitting there. But where? Exactly? This is where we need to enter the data into the machine for our first Work Offset, for Setup-1.

If you recall at the beginning of the book I discussed the purchase options for the machine, one of these was the Tormach® Passive Probe. We will use this now. How you initially set this tool up is covered admirably in the Tormach® Operator's Manual. I am going to assume for the moment, you have done this. It is not difficult and, for the moment, it is more important we maintain our momentum.

**PathPilot® Work Offsets automatic probing screen**
Copyright© screenshot courtesy of Tormach® Inc.

We will use this probe to measure exactly where the STOCK is in the vice jaws. The probe is placed into the spindle head – it has its own BT30 tool holder and a trailing lead which is plugged into the machine electrical cabinet. We will bring the sensing tip of the tool up against the STOCK from an x-axis direction, a y-axis direction, and a z-axis direction (this last one means moving down to touch the STOCK from above). Since it is plugged in to the computer it can send data back and the computer makes a note of the measurements. We don't have to carefully move the sensing tip ourselves – Tormach® have very kindly written some devilishly clever software macros which move the tip automatically to 'find' our STOCK edges. You just need to move into the general location. Press a button on the screen, and the computer takes the strain and does all the trickery for you. Brilliant.

So, for Setup-1, where shall we choose to make our origin? It turns out there are some standard suggestions which make complete sense so let's use those. We certainly shall not choose the origin to be anywhere on or inside the PART we are making. That's because the

PART is still 'inside' the STOCK and we have no way of sensing with a probe anything yet to do with it. Therefore, it is much more sensible if, for Setup-1, we choose the origin to be, say, top left rear corner of the STOCK as it sits in the vice. I could choose top right. It makes no difference. I could choose front left, but I wouldn't. That's because normally the vice sits on the machine's table with the fixed jaw at the rear and the moveable jaw at the front. Much better to choose a fixed jaw for a reference because it is never going to move. Just to cover all bases, I could have chosen bottom left rear. Yes, that's true, if I had sat the metal STOCK on a set of parallels in the vice, I could get my sensing probe to detect the top of the parallel and that would be the same as the bottom of the STOCK. We will come back to that in a moment. For the time being, for Setup-1, let's stick with top left rear corner of the STOCK and make that our 'manufacturing' origin.

When we first power up the Tormach® 770MX from fully off, the machine's controller, PathPilot®, does not have the feintest clue where all the axes are. It's like waking up from a deep sleep in a hotel room not knowing what country you are in. Yes, as bad as that. So, on the touchscreen we press reset and then the reference button for each of the three axes (there are no actual buttons – these are all on the touchscreen). The servo motor for each axis then drives to its mechanical limit which is a 'stop' microswitch and the 'brain' of PathPilot® figures out where it is again. At this stage, the machine only knows its own description of 3D space, it's working envelope if you will. Whenever the moving parts of the machine travel, from now on, the machine will know where they are, but encoded in its own numbers system, something we need not know anything about.

When we probe the points on the STOCK using our electronic probe, we are effectively telling the machine where we want to think of the axes as zero value. Consider if you were probing the left edge of the STOCK, you would be telling the machine where X=0. It is not zero to the machine, it still 'knows' where it is in its own world of numbers, but it will assign that position the value zero for us. And likewise for the other two axes. If you already run a manual mill or lathe which has digital read-outs (DRO), this is akin to zeroing an axis.

It soon becomes clear this is a useful system and we will soon be wanting more origins we can store in its brain. One won't be enough, so we need to give each one a label so they don't get confused. Each time we design a new setup, we will name the origin for it the work offset. By convention, the first one is named the G54 offset.

### 7.2.1 G54 work offset

G-code is a funny thing and when you are learning from scratch what everything means it gets awfully confusing because some G-codes are commands, like 'move in X and Y somewhere', while others are statements like 'I want you to think in metric please', and others are data stores where you save 'places in 3D space' like work offsets. It is not that complicated but when you are new to the subject it can appear so, especially if you are trying to teach yourself. Some specialised G-codes seem to be interpreted differently by different manufacturers; the main ones are standard of course. Later in the book, I will generate a huge list of everything I can think of which is relevant to the Tormach® 770MX. But, right now, let's crack on with G54.

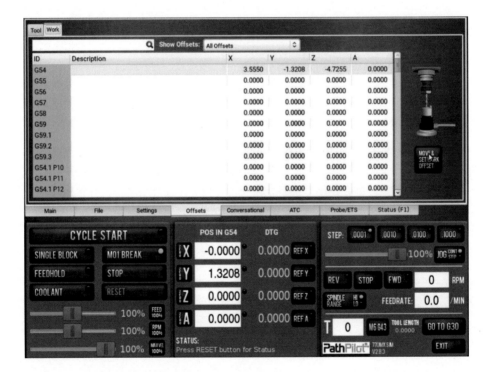

**The Work Offsets screen starting with G54**
Copyright© screenshot courtesy of Tormach® Inc.

There are up to 500 data locations in computer memory store where we can store 'origins', but right now we don't need that many, and it gets complicated as to how they are labelled after the first six. Let's just allow ourselves those six for the time being - they are called G54, G55, G56, G57, G58, G59. These are all different workspaces, or work offsets, and on the PathPilot® control screen, on the tab called OFFSETS, you will find all of them listed. Think of them as different origins – places in 3D space where we are going to want the controller to think of the values of X, Y and Z as zero, for our various setups.

For our Setup-1, we are going to use G54 offset for our origin. What I expressly mean by that is I want the controller to think of the top left rear corner of my piece of STOCK as a place in 3D space where X, Y and Z are all zero value. And I am going to label the space as G54 space. I could simply type the actual values for X, Y and Z into the tables, but that would involve me knowing 'machine space' numbers which I clearly do not. Because I have no real idea of where that corner of the STOCK is, in terms of the computer's own machine numbers, it is surely much cleverer to use the sensing probe to fill in the numbers for us. And that is exactly what happens. However, when we probe the STOCK automatically it will send the data to the offsets table for the 'offset' mode the machine is currently in. On the PathPilot® screen, we can read the current mode, or workspace. In the screenshot above, it is G54. (Normally it would also have this information in the Status line, but you can see I forgot to press reset before taking this screenshot. Sorry.)

If it is not the workspace we want, so say the Status line said G55, and we wanted to enter our probe data into G54, then we must change it first. The probe will *always* send its data to the current work offset. To change the current work offset, we must click on the first tab called MAIN, then type into the box named MDI (manual data input). In this case, we would type G54 and 'enter' and immediately you have commanded the machine to think referenced to the G54 origin.

Once we have probed the STOCK from all three directions in G54 Workspace, then that is where our machine will continue to think the origin is, just so long as the Status of the machine remains in G54 mode. Gosh I made a bit of a meal of that, but it's so important to get this right; if you confuse Workspace origins, between what you tell your machine, and what you plan in the CAM software, you will very quickly crash the machine.

### 7.2.2 Flip for the second setup

Later, when we have made the top half of our part we will 'flip' it over in the vice so we can do the second setup and complete it. And we will choose a new origin for Setup-2 and use G55 workspace. It is extremely important the origin we choose in the CAM calculations from the Manufacturing space of Fusion 360 is identical to the origin we sense on the MATERIAL in the vice. Notice I wrote MATERIAL this time, and not STOCK or PART. We need to choose our references very carefully and be precise what we mean. For Setup-2, we will be machining the flip side of the part we started in Setup-1.

And in doing so, it is likely we shall hold the freshly machined end of the PART in the vice, upside down from Setup-1. This means the bit the vice is grabbing hold of is PART, not STOCK, since we machined the stock away.

Ideally then, in the CAM settings for Setup-2, the origin we choose would be based on the PART, not the STOCK. And we would need to find a suitable point on the PART which we can easily probe. If you are using Fusion 360 for your CAM, it is best to just play around with the manufacturing space and learn how to pinpoint where your choice of origin should be, within the part or the stock, and if you find this tricky, YouTube is the way to go.

The Y axis is easy to sort out. Let's use the fixed jaw of the vice again. And this time it is clamped to the rear of the PART. On the machine we don't even need to probe it – the vice hasn't moved anywhere, so we could simply go to the work offset tables tab, and whatever is in the Y column for G54, copy the same number into the G55 row.

Now let's think about where the Z origin will be, in our CAM Setup-2. We could use the top of the STOCK again, but STOCK is not accurately cut to size and so if we did this, we would end up with our finished part being a thickness that we didn't really know because, after flipping the part over, we had referenced in from an unknown quantity. That's no good. So, for Setup-2, it would be more useful if we sat the flipped part on a set of parallels and then probed the top of those parallels with our sensing probe and call that Z=0 (for G55). And in the CAM setup, we would make sure the origin, as far as Z is concerned, was the bottom of the PART for Setup-2. Don't forget, when we started making the part during Setup-1, the top of the PART is what I am now calling the bottom! There is a serious point here; it can get quite confusing, so do make notes, and describe to yourself each setup, what work offset reference you are using, what you are calling the top and bottom of your part (and stick to that choice religiously) and maybe some explanations as to why you made those choices, to remind yourself later. Certainly, in the Fusion 360 CAM software, it is useful to change the names of the setups to your own chosen names to help reduce any confusion. For instance, Fusion 360 might name the first setup 'setup 1' but you might wish to call it 'threaded hole end', or some such suitable description.

What about the X reference in G55 for our Setup-2? Well, once we have completed both setups, we are hoping for the sides of our part to line up smoothly enough so we can't see the join. That should easily be achievable in the Y axis since the fixed jaw of the vice won't move. But in the case of the X axis, we will need to probe for a surface in the X direction. There is absolutely no point at all in probing the STOCK for the X axis in Setup-2. Nothing would line up in the X axis from Setup-1 and we will have made a shoddy part. Instead, we must probe the left side of our PART that was machined in Setup-1 and call that X=0 in G55 workspace. The trouble is it might be a bit tricky for our probe to reach the PART because now we have flipped it, the unmachined STOCK facing upwards is wider and will tend to shield the probe's access. We need to find a way to reach in underneath the overhang. That's not possible, so the next best thing is to place (and hold) a known spacer flush against the side of the PART. When the Tormach® goes through its sensing routine, if the spacer is held to the left of the part, the value PathPilot® assigns to the work offset table for X will be less than it should be by an amount equal to the thickness of the spacer. So, grab a calculator, add this to the number in the table and type in the new value.

Always be aware, the X axis increases from left to right, the Y axis increases from front to rear, and the Z axis increases going upwards. That is always true and worth getting your head round. In the example above, as the probe was moving from left to right trying to bump into the left side of the PART, it came across the added spacer first and had not 'increased in X' enough because it had not reached the PART yet. That's why we had to add on the spacer value. Had we been probing this from the right side instead, then the opposite would be true.

I hope that has added to your understanding of the work offset system. The machine now knows where the material is. We now need to find a way of telling it where the cutting tool is.

## 7.3 Tool offsets

Every time we put a tool into a BT30 tool holder and tighten it up, we have no idea how long it is, nor how far up into the collet it has slid. We need to measure it. Then we will label it (give it a name that is easy for humans to understand) and then we will assign it a tool number (because computers prefer numbers).

And we will end up with a lot of different tools, all pre-set and measured in their respective holders, and the data can be stored in a tool library. Because this is a milling machine, all cutting tools have two parameters which the machine's computer needs to know, the diameter and the length. And by length, what we need to know is how far the tool protrudes from the spindle, so the control software can always be certain of the tool tip position in the Z axis.

For our purposes, let's assume we have bought a brand-new end mill, that has zero wear, and what diameter the manufacturer has etched on its shank, we will trust to be right. To enter the tool into the machine's tool library, open the Offset tab on the PathPilot® screen, click on Tool Offsets, and there you will find the tool library. Start at the top, tool number 1, give it a name, and enter the diameter in millimetres. But what about the column marked length? How shall we measure that? We could 'jog' the machine downwards to bring the tool tip near to a reference like a 3-2-1 block sitting on the table, test for a sliding fit, and then enter the data into the tool table, bearing in mind how tall the 3-2-1 block is. There are various ways to do this which have been around for years. They are all well documented in other reputable resources but, to me, they are the metaphoric 'SMS texting' I mentioned before, when I would prefer to be using the latest messaging 'app' instead.

That is why I chose to buy the Tormach® ETS or 'Electronic Tool Setter'. It is a superb piece of equipment, bolts securely to the table and, once set up, is fully automatic. By that, I mean, rather like the passive probe sending data straight to the work offset table in PathPilot®, this clever tool sends the length data for the tool straight to the tool library. (And just like the passive probe, initial setting up of the ETS probe is well documented in the Operator's Manual.)

Once you have secured the ETS tool probe to the table in a suitable place out of the way of the vice, you enter its lateral position into the PathPilot® software (it is user friendly and explains how on screen) and so whenever you begin with a brand-new tool, entering it into the PathPilot® tool library is a non-event. You simply enter the tool number on the screen, then press 'set tool length'. And it does it quickly, accurately and without fuss. And crucially, without having to think much about it.

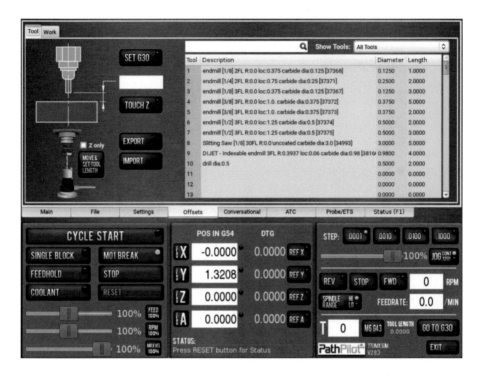

**Tool Offset data is entered automatically using the ETS probe**
Copyright© screenshot courtesy of Tormach® Inc.

Here's how: let's say I just inserted a brand-new 6mm drill into a BT30 tool holder with a 6mm collet and tightened it securely. I want to call it tool number 72. I must now manually insert the tool holder into the spindle, ensuring the mating surface (the taper) is clean, so it is securely held and repeatable, and then I type the number 72 in the box marked 'T' for tool, on the screen. PathPilot® now knows tool 72 is in the machine. But it knows nothing else at all about this tool yet. Next, I must navigate to the tool library page and on the row marked 72, I will type in the description box something along the lines of '6mm drill', hit enter and the cursor will jump to the diameter box. Now I must type in '6' and enter. So far, so good, these are all things I knew. It now asks for the length. This is something I don't know, so I click the 'Move and Set

Tool Length' button. The mill table bursts into life and moves to a position so the spindle lines up vertically above the ETS probe (because we previously set this up) and then slowly the spindle head brings the tool down until it touches the probe. It immediately retracts a tiny amount, does a second more gentle touch for greater accuracy, then swiftly retracts the tool out of harm's way. Within less than a second, you will notice the length data for this tool now appears in the library. Fantastic and easy.

### 7.3.1  G43 code

Whenever you put a tool into the spindle, you must tell PathPilot® the number of the tool you want to call it. Otherwise, it wouldn't know where to store the data. There is a strange G-code called a G43. And this is also one of those odd ones which doesn't seem to do much, but it is always popping up in the programs which your CAM software produces, at the point where you start to use a new tool. If the code calls for G43 mode, it is telling the machine it must use the tool length data in a certain way. Basically, the Z axis increases in value going upwards, remember, so any tool pointing downwards from the spindle protrudes in a negative Z direction. The code G43 merely tells the machine we measured the length of our tools in the sense their length values are positive numbers. Since there are different ways for machine controllers to work, this code must be used, but please don't ever worry about G43 again. All you need to know is, we will always get our machine to use G43 mode, and the length data in the tool library for each tool will always be positive numbers.

### 7.3.2  Tool libraries (real and not-so-real)

Notice I wrote tool libraries but there is only one tool library in PathPilot®. I make the distinction that the Tormach® 770MX tool library knows about all the tools you have set, measured, and entered and it only cares about length and diameter of each one. Whereas, in the 'manufacturing space' (the CAM environment) of Fusion 360, you can have as many tool libraries as you like! And the tools in these libraries may not even exist. You will need tools in this 'make-believe' environment to proceed with the calculations that CAM performs, to create G-code programs to manufacture your parts. You can have a main library of all your tools, and sub libraries for each part if you wish, and they are all stored on the cloud, which is exceedingly useful. The Fusion 360 tool libraries store much more data than just length and diameter. In fact, they don't even know the true length of a tool, even if it did exist, because that data is specific to your machine and how it holds that tool. What tool data it does store, though, is crucial for the CAM calculations the software needs to do. This is data such as tool type (end mill, face mill, drill, tap), speed of rotation (rpm), expected depth of cut, width of cut etc. and you can even store manufacturers' part numbers for easy replacement. And, of course, a tool number.

Strangely, sharing your tool data between Tormach® and Fusion 360 is nontrivial. I have not had much success with this and, so far, have tended to resort to just typing my tool descriptions into the PathPilot® screen manually. It doesn't take too long. And of course, once you design a part and then design all the CAM for that part within Fusion 360, you are going to need to have real tools in your machine which correspond to the tools you told Fusion 360 about for the CAM calculations.

It is quite easy to cause confusion between tool libraries on the cloud and the actual PathPilot® tool library which exists on your Tormach®. It is vital to ensure the tool numbers correspond exactly. That sounds obvious but, believe me, it is so easy to get wrong, with disastrous results. After making some mistakes in this area, whenever I place a tool in a holder in the machine, I now always physically give it a sticky label with its tool number, and enter it into the PathPilot® machine library, and pedantically always make a point of entering the same tool into my Fusion 360 tool library on the cloud. (In Fusion 360, you do not need to open a drawing, simply click on 'Manufacturing' workspace, and the tool library tab is there.)

## 7.4  G30 code

The G30 code is an interesting and useful one to know. Think of G30 as an instruction to move parts of the mill to a safe place. Imagine you have secured the vice to the table in a roughly central position and there is some STOCK in it. That's something to think of as a potential obstacle to a cutting tool sticking out of the spindle. Things appear potentially hazardous when you realise the automatic tool changer, which is situated to the left side of the machine, is not that high up. Whenever a tool is loaded from the automatic tool changer, particularly if it's quite long, as the carousel retracts away the tool is left 'dangling' quite low down compared to the vice. If there was a move of the table at this stage, it is quite possible there could be a crash of the tool into the STOCK in the vice, or even the vice itself.

So, there is a place in space, in X, Y, and Z, where the table can move to, and the spindle can move up to, which we could think of as a safe place, thus moving the STOCK in the vice away from the automatic tool changer, and the spindle high up as we need so the tools don't dangle too low and risk a crash. This can be anywhere we choose. And we get to program that place in space and label it G30. I currently set my G30 position with the table moved over to the far right, and the spindle near its upper limit. We jog to this position manually using the 'jog' handle or keyboard shortcuts, then we program the position using a button on the tool library page.

On the front screen, at bottom right just above the EXIT button, is the G30 'command' button. Whenever we click that button, the machine will go there. Usefully, it does the Z movement first, so withdraws any tool upwards safely out of harm's way before starting any sideways translation. Also, we must have previously gone to the Settings tab and made sure the box marked 'G30/M998 Move in Z only' is NOT ticked. I don't have knowledge of why ticking this box might ever be particularly useful.

When Fusion 360 generates a G-code program, it commands a G30 in the initial set up before any machining is started. If we have made sure our chosen G30 positions the vice and the work well to the right, out of the way, we can be certain a call to fetch a tool isn't going to be a concern for us. Secondly, and rather importantly, Tormach® have implemented a G30 move after every automatic tool change occurs. (We won't see this in the program code – instead it is within PathPilot®'s 'macro' for a tool change invisible to us as users.) If our G30 is set higher than the tool change level, then our newly selected tool will always retract up and out of the way before the next command to move towards the work takes effect. Interestingly, an earlier version of PathPilot® did not do this and it was possible to fetch a long tool from the automatic tool changer and crash straight into the work because the Z 'clearance' height of the next cutting sequence came to the machine's control too late. That's long since fixed and if we have a safe G30 height, we can be sure the next tool will always go up to G30 height, move across to the next cutting position and, only then, descend to the cutting height.

In summary, after an automatic tool change, the tool will always be at our G30 height so long as we have selected a G30 height which is higher than the tool changer, so I strongly recommend setting this up accordingly.

This problem does not manifest itself if we load a tool manually into the spindle because we are bound to move the spindle up quite high to allow ourself the physical space to load the tool. If I wanted to perform a manual tool change and there was not the space I needed, I would press the 'Go to G30' button and that would give me the access I needed without even having to think.

As an aside, the code which Fusion 360 produces also puts a G30 command right at the end of a program. I have seen some opinion on YouTube suggesting programming our G30, so our newly finished part is thrust to 'centre stage' for ease of access, and ease of removal from the vice. That's great advice for large machines which don't have their automatic tool changer so close to the action, but my advice is always have a 'safe space' G30 on the Tormach® 770MX. There is another programmable space available in G-code, labelled G28. Whilst there is no direct button for programming this location, it is instead performed by entering the instruction G28.1 into the MDI box and pressing 'enter'. So, this might provide an opportunity to 'push' a finished part to the front of the machine for easy retrieval after manufacture if we really felt the need. However, I tried this and found, unlike with the G30 position which works well, my PathPilot® controller forgets its G28 location whenever the power is turned off. In any case, does it really matter where the table ends up when our part is finished? Absolutely not, you can simply move the table to a more convenient position to 'unload' the fresh part using keyboard 'shortcuts.'


7.5  Conversational programming

I keep mentioning Fusion 360 because I think it is fabulous, and you will really enjoy the CAM side of things when you start to use that. But there will always be times when we simply need to face off a piece of material, or machine down a piece to a size, or drill a hole, or something basic such as these, where we don't need a full CAM setup.

Tormach® have cleverly included a tab in PathPilot® called 'Conversational' which is some smart macro programming they have written for us to perform ordinary functions like facing off a top surface, cutting a hole, and many more interesting operations. We do not have to write any G-code. We don't have to resort to any outside CAM program. We simply go through a set up routine – basically answering a few questions on the screen about what we want and where we want it – and it generates the code for us. It's good, but it has its limitations, of course. Creating actual G-code with CAM software is still going to be the better option for designing and making parts but, for a little bit of machining here and there, it will serve us well. And that's exactly what we need right now, to make our first cuts. Let me talk you through it.

Place a piece of rough stock into the vice and probe it with the Passive Probe in G54 space so you have, say for example, the top left rear corner of the STOCK as your G54 origin. You will select the FACE operation and immediately become aware of how intuitive it is. You simply tell the machine, in the Conversational page for FACING, all the data it needs. And then it converts that to a 'mini' G-code program. Interestingly, there is a choice of 'spiral' facing or 'rectangular' facing and if you click these options, you get a clear diagram on the touchscreen of what those mean.

Looking at the diagram on the 'conversational' tab shown, if, say, we had a piece of STOCK in the vice which was 100mm long (in the X direction) and 30mm wide (in the Y direction), and we had probed the top left rear corner as our G54 Offset, then we know that X starts at zero (because we probed that to be the origin), and ends at 100. Likewise, Y starts at zero and ends at minus 30 (a minus sign is required since the diagram on the screen clearly shows the 'Y end' is nearer the front of the table than the 'Y start' and so the machine is going in an increasingly negative Y direction to do this job).

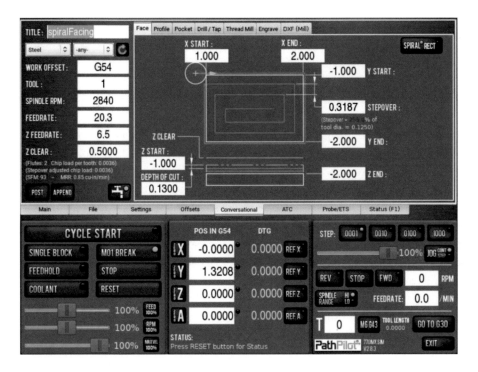

**Conversational Programming with PathPilot®**
Copyright© screenshot courtesy of Tormach® Inc.

So, let's say we wanted to face this piece with a 6mm end mill. Not a great choice, but for the sake of argument, let's say that's all we had. Two things come to mind. Firstly, we really don't want to get any burrs at the edges so it might be best if the cutting tool passes a bit beyond the edges. For this reason, we will set 'X start' at minus 1 (1mm left of the STOCK) and 'X end' at 101 (1mm past the right edge of the STOCK). That way we are making the conversational mode think the metal is 102mm long and we will be machining just past the edges. A similar argument makes our choices for 'Y start' to be 1, and 'Y end' to be minus 31, a total width of 32mm to machine when, in fact, only 30mm exists.

Next, we must consider 'stepover'. It is worth experimenting with this value to see the effect. For now, consider setting this to 5.5mm. In this way, as the cutter proceeds along a second parallel cut, it will overlap the previous cut by 0.5mm so avoiding any possibility of a burred edge. We always want the stepover to be less than the diameter of the tool for obvious reasons.

And it is worth playing around to get to grips with the whole G54 thing. What we must not forget is, once we have done the probing successfully, the top of the STOCK is Z=0. And if we wanted to shave off, let's say, 0.5mm we would be facing this top surface until we reached down as far as Z= -0.5 (for the sake of clarity, that is Z=minus 0.5mm). Going downwards is negative Z direction. We have arbitrarily told the machine that Z is zero exactly where the top of the STOCK is, so anything lower than that must be negative.

Some books assume this is obvious. I don't think it is, so I hope you forgive my emphasis. Getting things wrong in the Z axis is a great way to crash your cutting tools into your work or even your vice. If you come from a FDM (fused deposition modelling) desktop 3D printing background (the sort where you lay down molten filament onto a print bed) and CNC manufacturing is new to you, then Z is still positive going upwards (that won't change), but you are used to starting at zero on the print bed and you always dealt with positive values of Z as your printed model gets taller. Negative values might seem odd at first but get used to them because this is how the CNC mill works. I suppose a good way to think about it is 3D printing is 'additive manufacturing' and your part builds up in positive Z space, whereas CNC machining is 'subtractive manufacturing' and you are cutting away material into negative Z space.

Just a reminder, the X axis increases from left to right. The Y axis increases from front to back. Again, this can also be confusing if we over think it. We are talking about the direction of the cutting tool which is doing the machining. When it moves from the left to the right as it cuts our STOCK, that is increasing values of X. It is as if the spindle moved to the right. It didn't, of course, because it's fixed. The design of the machine is such that we have a table which moves below a fixed head. When X increases the table moves to the left. A similar discussion can be had for the Y axis. But there is no need. Do not over think this. Think only of the motion the cutting tool makes over your WORK and you will not go wrong.

There are quite a few other functions you can perform within the Conversational Programming part of PathPilot®, and I recommend you grab some old pieces of aluminium and explore away. It is a powerful part of the machine and easy to use.

Oh, I forgot to mention, once you have filled in all the data blocks within the conversational milling page you are using, you simply press the 'POST' button and PathPilot® computes its own G-code for you, and places it into the machine's memory ready for you to try. You can give the file a name and keep it for later. Marvellous.

7.6   Post processor

I want to explain what a post processor is. I think it is another of those pieces of jargon which is not well explained and quite often publications, and online videos, assume you know what it means. When we were discussing choosing a machine, I made the point I was a huge fan of Fusion 360 and I thought it worthwhile buying a CNC machine for which Fusion 360 offered a post processor.

The post processor is the extra piece of software which we must download from the CAM software supplier and is a translator, to make sure the G-code Fusion 360 wants to create, is compatible with the Tormach® interpretation of it.

Note that Tormach® do not 'look after' the post processor; its upkeep is maintained by Autodesk, the company producing Fusion 360 CAM software. It is in their interest, of course, to be able to say they support the use of Tormach® machines. Pop over to the Autodesk website and download the post processor. Within your copy of Fusion 360, you will be able to import this into your software. When you start to explore the Manufacturing space, the first thing you will do is create a 'Setup' and that's when the software will ask you which machine you intend to use to make this part. Since you have downloaded and imported the post processor, you will find the Tormach® 770MX option and now you will be ready to create G-code for your machine.

7.7   Fusion 360 'manufacturing space'

This book's aim is not to teach Computer Aided Manufacturing (CAM). There are specific books for this. In fact, you will learn a lot merely by experimenting in the Manufacturing space of Fusion 360. So, go ahead, in the Design space, draw a simple part like a cuboid with some dimensions, maybe a hole in one end, nothing too taxing, and then go to the Manufacturing space and try to create the first setup for machining the first side of the part. Call it Setup-1 or give it a real label like 'top of my part'. In the setup, you'll need to specify your machine, the 770MX, and then your workspace environment. Let's use G54 and choose the origin somewhere on the STOCK. You will also have to specify how big the STOCK is, and whereabouts within the STOCK the PART sits. It is not too complicated, but it is more time consuming than you can possibly imagine. At this stage you will be sat at your computer a lot more than standing at the Tormach® machine, which might seem frustrating or wrong, but you're now in a position where you want to be able to create some G-code and see what it does on the machine, so this stage is vital. Getting used to the idea you might very well spend much more time working on the CAM software than watching your mill make parts is part of the learning process. I found exploring the CAM software very interesting and rewarding. I recommend you visit the YouTube channel of John Saunders at NYC CNC, where you can pick up a great deal of knowledge on using Fusion 360, particularly in the manufacturing space.

In keeping with the theme of this book, I want to continue to pick up on top tips I have gleaned from my personal journey. One of the first things you are going to want to accomplish is creating G-code from Fusion 360 CAM which performs the same functions you already did using the Conversational Programming within PathPilot® itself. It's all very well PathPilot® can do it for us, but to learn the art of CAM, we need to use simple examples and walk before we run. And in the long term, this will be more useful for our learning. That's my take on it, anyhow. The Conversational Programming system does not utilise the concept of your PART existing within your STOCK, so it's important to begin thinking about this aspect. Once we have initialised a setup within CAM, we will assign operations, such as 'face' or '2D contour', to enact within that setup. In each operation we will need to specify a tool from the library and give instructions about where and how much of the machining we need to do. Learn this by experimenting. Begin with a small tool library on the cloud within Fusion 360 and replicate it for real on the machine. We already discussed our Tormach® tool library within PathPilot®, but it is worth a quick look at the Fusion 360 aspect of tool library setup.

## 7.8 Tool library specific tips

Within Fusion 360, click on 'tool library', and use the 'plus' sign to add some tools to your 'cloud' tool library. By the way, I omitted to mention the tool number 99 is a special one on the Tormach® mill and is reserved for the passive probe we use for setting work offsets. Do not try to put any cutting tools in row 99. If you ever put your passive probe into the Tormach®, be sure to type 99 into the box on the screen, and immediately the mill 'knows' it has the probe. The tool library in the CAM software doesn't need to know about tool 99. (It is possible in more advanced programming, when you might be coding probing routines within a manufacturing process, using a wireless probe which can be stored in a tool changer, you would then need details of the probe tool in the library, but I don't have that level of knowledge I'm afraid.)

Sitting at your computer, as soon as you create a tool in the library you will give it a number, but the tool is just a figment of your imagination at this stage. Don't forget, as soon as you create the tool for real, and set it up as part of your actual library on the mill, you must assign the same tool number and, to ensure you don't muddle things up, stick a label or at least a 'marker pen' number on the tool holder. I have a label maker made by BROTHER which produces self-adhesive labels, and these hold up well to all the coolant swishing around. That's something to bear in mind actually; they will get splashed a lot. You need the tool number labelling to be resilient.

It is so important the tool you think you are using, and the actual one which is in the spindle, are the same number, and the same tool, for obvious reasons. And the length data for the tool must be correct in the PathPilot® tool table. It makes sense the tool library in Fusion 360 should mirror your tool table in the machine. Of course, for planning purposes, you may well have tools in your CAM library which you have not bought yet. On the other hand, it is less useful if you have tools in your machine which your CAM software has no knowledge of.

Initially when I was going through this learning process, I didn't yet have the labelling system sorted and I 'thought I remembered' which tool number I had assigned to the tool I was using. It turned out I had made an error and although the length data for my tools was all correctly recorded, the machine tried to face off the top of the STOCK with one of my end mills whilst using the length data for the tool number of another of my end mills which I had told the machine it was using. The upshot of this was my first use of the Emergency STOP switch in desperation, when my machine started facing off the top of my vice jaw.

That was a very early and salutary lesson for me to label my tools carefully and accurately, and not try to rely on remembering. There are a few lessons like this which have no equivalence in the manual milling world and, because of this, come at you as a complete surprise. This was totally my fault. I didn't think I would do it. But I still did. Feel free to learn from my errors.

Another thing worth mentioning – it is not obvious and sort of assumed we will figure it out – once we have put a tool into the spindle, entered the tool number in the box on the screen, and measured its length, the tool is ready to use. You can take it out of the machine and sit it in a rack but, if you have space in your automatic tool changer, you can pop it into the carousel straight away. It's so easy; select the ATC tab on the PathPilot® screen and hit the 'Store Current Tool' button.

And here is the point. The representation of the tool changer carousel on the PathPilot® screen (in this case with ten positions for the 770MX machine) now shows your tool at one of the locations in the carousel. You never need to know what carousel location that is. It has no number that you need to know. The machine clearly knows where that tool is stored but all you get to see is the tool number. If I had stored an end mill labelled tool number 23, and then popped a drill labelled tool number 67 into the machine, and hit 'Store Current Tool', the machine would place the drill next to the end mill in the carousel and the numbers 23 and 67 would clearly show these were the tools stored. And it will not forget even if you power down the machine. This is non-volatile memory within the ATC control circuitry and works reliably.

Summarising all of this, describe your tools accurately, number them reliably, always make sure PathPilot® has the length data for all your active tools and, finally, keep a mirror of your actual PathPilot® tool library in your Fusion 360 tool library on the cloud. It is much more realistic and sensible if you can call on tools within your CAM environment which are actual tools in real life.


7.9  G-code preamble

Now we have created some G-code for simple operations, we can have a look at some of the commands. It is surprisingly human-readable and with a de-code of some of the more common commands you can soon start to get a feel for what is happening.

Before the core part of any code which gets the machine cutting, there is always a sort of preamble of code to place the machine into the correct modes. A very useful piece of free software is something called Visual Studio. I highly recommend you download this onto your computer. When the post processor within Fusion 360 creates G-code, it is stored as a file type ending in '.nc' and if you set your file type associations so that .nc files open in Visual Studio, that is a useful way to read, and indeed edit, your code.

```
       File   Edit   Selection   View   Go   Run   Terminal   Help

       ≡ OP1 G54 socket end.nc  ×

       C: > Users > iiiii > Dropbox > TORMACH > Fusion 360 > Posts > example post > ≡
        1    %
        2    (OP1 G54 socket end)
        3    (post version 43429)
        4    (post modified 2021-09-16 080638)
        5    (Machine)
        6    (  vendor Tormach)
        7    (  model 770MX)
        8    (  description Tormach 770MX)
        9    (T2   D=8. CR=0.  - ZMIN=-14.7 - flat end mill)
       10    (T4   D=4. CR=0.  - ZMIN=-15.  - flat end mill)
       11    (T6   D=6. CR=0.  - ZMIN=-12.  - flat end mill)
       12    (T7   D=40. CR=0. - ZMIN=-0.5 - face mill)
       13    (T14  D=3. CR=0.  - ZMIN=-12.  - flat end mill)
       14    (T21  D=6. CR=3.  - ZMIN=-22.5 - ball end mill)
       15    (T32  D=6. CR=0. TAPER=118deg - ZMIN=-13.703 - drill)
       16    (T39  D=2. CR=0. TAPER=118deg - ZMIN=-12.601 - drill)
       17    (T42  D=6. CR=0. TAPER=90deg - ZMIN=-3. - spot drill)
       18    (T70  D=6. CR=0.75 - ZMIN=-7.298 - slot mill)
       19    G90 G54 G64 G50 G17 G40 G80 G94 G91.1 G49
       20    G21 (Metric)
       21    G30
       22
       23    N10(Face3)
       24    T7 G43 H7 M6
       25    S2200 M3 M8
       26    G54
       27    G0 X41.3 Y0.
       28    G0 Z18.
       29    G0 Z3.5
       30    G1 X41.292 Z3.249 F200.
       31    G1 X41.268 Z2.999
       32    G1 X41.229 Z2.75
       33    G1 X41.174 Z2.505
       34    G1 X41.104 Z2.264
       35    G1 X41.019 Z2.028
       36    G1 X40.919 Z1.797
```

**An example of the beginning of some G-code**
Copyright© screenshot courtesy of Visual Studio

As an example, I have taken a snapshot of the beginning of a very simple piece of code as it is portrayed within Visual Studio. Let's go through some of the initialisation set up code. As you can see, it starts with a percent % sign to say: 'what follows is G-code'. Everything in brackets is comment, purely to help us understand some basic information about the upcoming code. Usefully, Visual Studio colour codes these bits green.

This is the first 35 lines of some code which I created using Fusion 360. In the manufacturing space when I did this, the first setup started with a facing operation. Importantly, I renamed that setup as 'OP1 G54 socket end'. I did that because those things mean something to me, so immediately I can read that line and know what I was thinking at the time. I also remind myself I am working in G54 work offset for this setup.

The Fusion 360 post processor has helpfully recorded which machine this code is written for, namely a Tormach® 770MX. And then it has listed all the tools I have used in this setup. That is super useful when it comes to making sure you have put all the tools you need into the automatic tool changer. Now you can see why it is so important your tool numbers within Fusion 360's library correlate directly with your real tool numbers in the PathPilot® tool library.

And then we come to lines 19, 20, and 21 which are written in blue. These are not comments; they do something. I shall look at each in turn.

### 7.9.1 G90 command

This places the machine into 'absolute distance mode', meaning that all instructions to move in any axis must be based on the actual co-ordinate with reference to the chosen origin. That sounds quite complex but it's not. An example helps; if I wanted to drill a hole 25mm in from the left edge of my part, and I had already set that edge to be my X origin in the G54 Workspace, then the X co-ordinate for the hole is 25. It doesn't matter from where we start, if we tell the machine to go to X=25, it will. This is G90 mode. There is another mode, called G91, which is the 'incremental distance mode' and allows code to move in increments relative to where you already are. That's great, but we hardly need worry about any of this. Usually, you will want to be in G90 mode and luckily, whether you are using code generated by Fusion 360 posts, or code generated by the conversational mode within Tormach®'s

PathPilot®, this is already set up for you, and that is why this G90 is showing in the example above. I didn't put it there; Fusion 360 did.

### 7.9.2   G54 command

As already discussed, this is the workspace to define our origin, which should match the origin in the manufacturing workspace in the CAM setup. But interestingly, you will notice as well as G54 being called for in line 19, it is also called again in line 26. Why? I don't honestly know why. But what I have gleaned is later, when you get a bit more advanced, you might wish to run more than one setup in a single program, in other words do a bit of work on one part maybe in Vice 1 and then do a bit of work on another part in Vice 2. Each fixturing setup (vice) will have its own designated work offset, so maybe G54 and G55. It is common practice for the second setup to be the 'flipped' side of the first. That way, every time you run this program you will end up making ONE part. When you combine setups in this way, you can optimise the tool changes. This is getting to sound advanced but stay with me. Let's say both the operations needed to be faced off using the same tool. There is no point in facing off the part in the first setup, putting the tool away, only to come back later needing to use it again for the second setup. Fusion 360 can compute this automatically for you if you choose to post process two or more setups into one program, so improving the efficiency of the manufacturing. It is very clever, as you might expect. The important thing to remember is, by doing this, your machine will be switching back and forth between different workspaces, i.e., between G54 and G55 modes in this example. Therefore, it is vital the code generated by the post processor re-establishes the correct workspace mode *every* time a new operation is called. In my example, line 26 in the code does this.

### 7.9.3   G64 command

This is called the 'blended path control' mode and is something to do with smoothing when you are asking the machine to cut round curves and corners and is quite complex. My advice is not to concern yourself about this for the moment. When you start to fiddle about with each operation within Fusion 360, you will notice there are parameters for smoothing and I am sure this is all accounted for by the software. Trial and error, as so often the case, is the way to learn what is happening when you make changes to these parameters. But for now, it is OK to accept we need the G64 command in there. The opposite of G64 is the G61 mode which is called 'exact path control' mode. Let's not worry about this for the moment.

### 7.9.4  G50 command

This is some sort of scaling command and I think it just resets all scaling to unity. I can't find much about the G50 command, and it hasn't hampered my learning curve as I gain experience with the Tormach®. I am not going to worry about this one either.

### 7.9.5  G17 command

This is about 'circular interpolation' and selecting the axes in which you intend to do this. The corresponding commands for what I would term non-normal plane selections are G18 and G19. For normal X-Y axes in the normal sense, it is easy to visualise the cutting tool moving in X and Y to cut in a circular sense, like making a protruding circular boss from a piece of square stock. Unless you intend getting into complicated motions where you'd want to cut circles in other axis pairs, I cannot think you will be using anything other than G17 for a very long time. But it is important the command is there; the machine must know which mode you are in, obviously.

### 7.9.6  G40 command

The commands G40, G41 and G42 are 'cutter compensation' commands. G40 says don't have any compensation and the other two add compensation, left or right. Think of it this way; if you wanted to cut a hole with a drill, then you would want to move the work in X and Y so the drill lined up exactly where you wanted the hole. You wouldn't want any offset – you want the exact centre of the drill (which is the exact centre of the spindle) to be at the X and Y co-ordinates of the intended hole. This is G40. Now, consider if you wished to mill some material to end up with a square boss which measured, say 25mm x 25mm, then the actual movement of the machine table in X and Y would depend on what diameter cutting tool you had chosen. So, if you had chosen an 8mm end mill, the machine needs to offset by the radius of the tool and the movement of the table in X and Y would be a square measuring 33mm x 33mm. Depending which way round you went, the compensation would be one or other side. (Actually, if you think about it, you could make the motion a 33mm x 33mm square with rounded corners and this is an option in more advanced programs depending on whether a sharp-edged corner is important to you or not.)

But here's the thing: you won't ever see any G41 or G42 codes in your programs which are created by a CAM post processor. This is because it is easy for a modern piece of software to calculate the exact paths in X and Y based on a knowledge of the radius of the tool in use. In short, it is done for you. The machine will run in G40 mode – cutter compensation OFF – and the post processor will look after the whole issue, guiding the table motion and making allowance for the tool diameter. Another thing not to worry about.

### 7.9.7  G80 command

Some special operations in a CNC machine are referred to as 'Canned Cycles'. These are processes such as drilling a hole (and there are various ways to do this; drill it in one go, take pecks at it retracting the drill after every peck, take pecks but only partially retracting after each peck), boring a hole, and tapping a thread inside a hole. The reason they are termed 'Canned Cycles' is they are a sort of shorthand the machine controller understands describing the operation very concisely. The commands for these have different numbers such as G81 and G84, to give two simple examples, and the important thing to note is when they are finished, the 'Canned Cycle' mode must be cancelled. The G80 command does just that and puts the machine back into normal mode ready to accept standard motion commands. Having a G80 command in the preamble at the beginning of every program is something the post processor seems to do as a safeguard.

### 7.9.8  G94 command

Let's talk about 'Feed Rates'. Although I shall discuss 'feeds and speeds' in a later chapter, this mode is merely setting the machine to operate in one of three manners. There are three commands which correspond to these: G93 sets the machine to expect feed rate commands in a 'how many times per minute' format. This is something they use on 5 axis machines and is beyond the realms of this book. G94, on the other hand, is very much the mode we wish to operate in and corresponds to the machine expecting feed rates in the format of distance of travel per minute. The standard feed rate in North America is expressed in 'inches per minute'. The standard feed rate in the UK and Europe is expressed in 'millimetres per minute.' Finally, the G95 mode expresses feed rate in terms of distance per revolution – you can see how this would become relevant when correlating the downward drive of a tap as it revolves to cut a thread.

Luckily, we will just remain in G94 mode and if there is any tapping to do, we will let the 'tapping cycle', the post processor produces, take the strain. Once again, we needn't worry ourselves with this too much.

7.9.9 G91.1 command

Soon, we shall come across some common G-codes, such as commanding basic motion. These are the quintessential G-codes, beginning at G0 which is 'go somewhere as fast as you can' (used for getting the cutting tool to the position for cutting quickly, sometimes called 'rapids'). G1 is 'go somewhere at a chosen speed' (used for cutting at a specified feed rate). G2 and G3 are for motion in prescribed arcs, or circles. And these two are a bit cleverer. Think of it this way; rather than trying to compose the X and Y co-ordinates all the way around a circle – and of course the sky is the limit depending on the resolution you choose – the G2 and G3 commands create a shorthand for doing so, effectively by specifying the radius of arc and letting the controller get on with it, in simple terms.

Referring to the G90 command earlier, where we wished to operate always in the 'absolute distance mode', this is the one situation when the use of 'incremental distance mode' would be beneficial. In specifying a radius of arc to create a circular path, we don't want to have to work out where that radius is (in absolute terms), we just know it has a value and the controller can work out the rest. So, the G91.1 mode deliberately specifies, for all G2 and G3 commands, the additional data which allows it to construct the arc will be given as an incremental value. This is yet another of those very important preamble codes which you need never concern yourself with again. Thank goodness.

### 7.9.10  G49 command

This is a strange one. I think it gets placed in the preamble as another 'just in case' safeguard. Let me (try) to explain. We talked about 'tool offsets' and making sure the PathPilot® controller knows the lengths of all the tools in the library. To ensure the controller takes account of the 'tool offset' data for each tool, there is a command called G43. This says, 'whenever you use a tool, please make sure you take account of how long it is, by looking at the data in the table.' G49, conversely, cancels that instruction. Furthermore, whenever a new tool is called, the post processor will insert a G43 command at the same time so the use of the G49 really isn't that important, I would suggest, in modern-day code. I understand an M30 code (program end) also cancels the G43 mode, so it is all a bit overkill. Let's not worry about that one either. I think high-level programming in CAM has this covered.

### 7.9.11  G21 command

Line 20 has the command G21. This just tells the machine controller to work in metric units, millimetres. The other option would be G20 for our colleagues working in inches. The Tormach® 770MX remembers what it was using last time, even after full power down. Nevertheless, it is good policy to always specify this in the preamble. And the post processor does this for us. How does it know what we want? Simple, in Fusion 360, in the setup process within the CAM section, it asks us.

### 7.9.12  G30 command

On line 21, the command G30 just sends the machine to the 'safe space' we talked about a little earlier. It is a good place to start, just in case the next thing we wish to do is fetch a tool from the automatic tool changer, and we might have a combination of long tools extending low and work pieces in the vice extending high. By always specifying a G30 before we do anything else, we can avoid a potential crash.

That's the end of the G-code pre-amble.

## 7.10 The operational G-code

In the example program, you can see at line 23, the code starts to do something useful. This is where the setting up of the modes is complete, and it is now time to make some movement commands to cut some material. It starts at N10.

### 7.10.1 N10 operational code

This is purely a 'block number' and is a reference location to find where each separate operation 'begins and ends' within the larger program. These are conveniently inserted by Fusion 360 when we post process our code to create the program.

### 7.10.2 T7 G43 H7 M6 operational codes

This is a line of code to fetch a tool and get it ready to use. If you don't have the automatic tool changer option, then the PathPilot® controller asks you to insert the tool manually. If you do have the automatic tool changer option, then this command line will load the tool automatically assuming, of course, it knows the tool is in the carousel. If there is already another tool in the spindle, then the machine will store that tool first, if there is a space to do so. If there isn't, then the controller will ask you to remove that tool first. They have thought of everything.

So, in this example, it is tool number 7 we want to load. And when it is loaded, the G43 command instructs the machine to take account of 'tool offset' data held in the PathPilot® tool library. Bizarrely, the H7 command tells the G43 command which tool data we want to take account of. In other words, it is telling the controller to use the tool length data for the same tool we selected. You could, quite conceivably, have a different 'H' number to the 'T' number, but why this would ever be useful is beyond my comprehension. In my example, the numbers match and this seems eminently sensible to me. The M6 command simply means, 'fetch the tool.'

### 7.10.3  S2200 M3 M8 operational codes

This command line is next up, so let's break it down into its constituent parts; M3 says, 'switch on the spindle motor in a clockwise direction.' How fast? That's the speed in rpm after the letter 'S'. Finally, we will have some flood coolant, so the M8 command switches the pump on.

### 7.10.4  G54 command

The next line is just a reminder we want to do the next bit of machining in G54 workspace, so before we command any movement of any of the three axes, it is worth making sure we know which origin we are working to.

### 7.10.5  G0 command

And then the motion begins. There are some G0 commands, which are fast positioning commands (rapids), followed by G1 commands which need to know how fast you would like to move. These are the moves when the milling cutter will be doing work, so the feed rate is important. That's why a G1 command has an 'F' number at the end. In my example on line 30, you can see there is an F200 command, which simply means 200mm/minute. (Don't forget my little trick of dividing by 60, so knock a nought off and divide by 6. That becomes 20 divided by 6, so is just over 3mm per second, which is easy to visualise).

```
8907    G1 X-7.414 Y0.502
8908    G1 X-7.306 Y0.279
8909    G1 X-7.19 Y0.049
8910    G1 X-7.134 Y-0.057
8911    G1 X-7.053 Y-0.206
8912    G1 X-7.038 Y-0.236
8913    G1 X-7.023 Y-0.248
8914    G1 X-7.007 Y-0.243
8915    G1 Z3.
8916    G0 Z11.
8917    M5 M9
8918
8919    G30
8920    M30
8921    %
```

**An example of a program end**
Copyright© screenshot courtesy of Visual Studio

### 7.10.6 Commands to end a program

Let's look at a few of the commands you will commonly find at the end of a program. In the example shown, line 8917 has two commands which come just after some G1 and G0 motion, so presumably there was some material cutting going on, then some rapid motion up out of the way (this would be a Z clearance to move the tool safely above the work), then:

M5

This means, 'switch off the spindle motor'

M9

This means, 'switch off the coolant pump'

G30

You may recall the G30 command will move all the parts of the machine to our pre-determined 'safe space'. You don't have to do this, but Fusion 360 chooses to insert this right at the end, and I think it is very sensible. As I mentioned before, you could use G30 as a way of presenting the newly completed work to the operator by moving it towards the door. But the Tormach® 770MX is not large enough to cause you any concern about reaching the work. Much better to move stuff well away from the automatic tool changer end, as I described before.

M30

This is the code which means 'end of program'. And the '%' sign is part of that convention too.

Now that I have introduced you to the machine by way of 'First Cuts', I think you need to have a reference list of all the G code you might come across using the Tormach® and, of course, the PathPilot® controller. There are far more comprehensive books and internet resources for G-code that exist, but this basic summary for the Tormach® will certainly be handy close by.

Since the Tormach® PathPilot® control system is based upon the Linux CNC system, I would urge you to research that online, at your leisure, if you would like to extend your knowledge. You can find out more here... http://linuxcnc.org/

# Chapter 8    Useful PathPilot® G-code reference

Another essential online resource for building up your knowledge of G-code is Tormach®'s own information, which is not easy to find unless you search for it specifically. Type *Tormach® Supported G-codes Reference* into Google and you will find an amazing amount of information at https://Tormach®.com/supported-g-codes-reference

This chapter aims merely to record the main codes you will need to know. I hope in a handy format for quick reference, but I admit it's certainly not a comprehensive listing. Don't forget, the aim of this book has always been to get you 'up and running' with a useful and effective machine, in the quickest but safest way possible. The following list began as a cut and paste directly from Tormach® but was then redacted and simplified to only show those codes I feel will be most useful at this early stage.


8.1   G code

| | |
|---|---|
| G0 | Rapid positioning |
| G1 | Linear interpolation |
| G2 | Clockwise circular interpolation |
| G3 | Counter-clockwise circular interpolation |
| G4 | Dwell |
| G17 | Plane selection for circular interpolation |
| G20 | Imperial machine mode (inches) |
| G21 | Metric machine mode (millimetres) |
| G28 | Predefined position |
| G28.1 | Store G28 position |
| G30 | Predefined position (safe space?) |
| G40 | Cancel cutter radius compensation |
| G43 | Apply tool length offset |
| G49 | Cancel tool length offset |
| G53 | Move in absolute machine coordinate system |

| | |
|---|---|
| G54 | Work Offset 1 (also called 0 in Fusion 360 WCS system) |
| G55 | Work Offset 2 |
| G56 | Work Offset 3 |
| G57 | Work Offset 4 |
| G58 | Work Offset 5 |
| G59 | Work Offset 6 |
| G59.1 | Work Offset 7 |
| G59.2 | Work Offset 8 |
| G59.3 | Work Offset 9 |
| | |
| G61 | Exact Path Control |
| G64 | Blended Path Control |
| | |
| G73 | Canned cycle – peck drilling with small retract |
| G76 | Multi-pass threading cycle |
| G80 | Cancel motion mode (including canned cycles) |
| G81 | Canned cycle – drilling |
| G82 | Canned cycle – drilling with dwell |
| G83 | Canned cycle – peck drilling with full retract |
| G85 | Canned cycle – boring, no dwell, feed out |
| G86 | Canned cycle – boring, spindle stop, rapid out |
| G88 | Canned cycle – boring, spindle stop, manual out |
| G89 | Canned cycle – boring, dwell, feed out |
| | |
| G90 | Absolute distance mode |
| | |
| G94 | Normal feed mode, in millimetres (G21) per minute |

## 8.2 M code

| | |
|---|---|
| M0 | Program stop (needs cycle start to continue) |
| M1 | Optional program stop (needs cycle start to continue) |
| M2 | Program end |
| | |
| M3 | Rotate spindle clockwise |
| M4 | Rotate spindle counterclockwise |
| M5 | Stop spindle rotation |
| | |
| M6 | Tool change command |

| | |
|---|---|
| M7 | Coolant on (other types of coolant – separate channel) |
| M8 | Coolant on (flood coolant) |
| M9 | All coolant off |
| | |
| M10 | Unclamp automatic collet closer |
| M11 | Clamp automatic collet closer |
| M30 | Program end and rewind |

The following are used with the USB M-Code I/O Interface Kit option:

| | |
|---|---|
| M64 | Activate output relays |
| M65 | Deactivate output relays |
| M66 | Wait on an input |

## 8.3  Expanding on some finer points

I am going to pick out a few of the codes to add a little more information where I think it might be useful.

### 8.3.1  G4 command

G4 is a dwell command and makes the controller simply wait for the specified duration and then proceed with the subsequent code. It is as simple as that. I have used it to write a short program to give the spindle a gentle warm up, for cold mornings in my garage. Have a look at the code to understand how it works. It starts by switching on the spindle motor at a low speed first, for a minute, then increases the speed a little for another minute and so on until it has run for five minutes, sufficient I think to loosen up the bearings in a sympathetic way.

```
spindle warm up 5 mins.nc

T: > SMD G code files > SMD experimental > ≡ spindle warm up 5 mins.nc
 1  %
 2  (1003)
 3  (Machine)
 4  (  vendor Tormach)
 5  (  model 770MX)
 6  (  description Tormach 770MX)
 7  (  spindle warm up routine - 5 mins)
 8
 9  G90 G54 G64 G50 G17 G40 G80 G94 G91.1 G49
10  G21 (Metric)
11  G30
12
13  M64 P1 (green signal tower ON)
14  M65 P0 (red signal tower OFF)
15
16  G04 P1 (pause 1 sec)
17  S500 M3
18  G04 P60 (pause 60 secs)
19  S1000 M3
20  G04 P60 (pause 60 secs)
21  S1500 M3
22  G04 P60 (pause 60 secs)
23  S2250 M3
24  G04 P60 (pause 60 secs)
25  S3000 M3
26  G04 P60 (pause 60 secs)
27  M5
28
29  M65 P1 (green signal tower OFF)
30  M64 P0 (red signal tower ON)
31
32  G30
33  M30
34  %
```

**My simple G-code program for a spindle warm-up routine**
Copyright© screenshot courtesy of Visual Studio

### 8.3.2 M3 and M5 commands

You can see I have used the M3 command to start the motor. It is perfectly acceptable to send new M3 commands without stopping the motor in between. At the end, the motor is switched off with the M5 command. It is very simple, and it works, but whether it is doing any good or not I do not know. I probably ought to put some science into it and find out what is the best routine for the spindle bearings when they are cold. But anyway, this is still useful for learning some more coding.

When you insert a G4 command, you must add a 'P' value, and this represents how long you want to dwell (wait) in seconds. If you look up the G4 command elsewhere, you might find differing interpretations of this command, such as the 'P' value representing milliseconds instead of seconds, but Tormach® 'use' seconds. Hence, 'G04 P60' simply means 'wait sixty seconds.'

### 8.3.3 M64 and M65 commands

You will notice I have used these commands at line 13, 14 and then again at 29, 30. I was just messing around really, experimenting to see how the USB I/O device works. It's a simple little electronic box, but it seems to be reliable, simple to connect, and I am already finding it useful. I purchased a signal tower from RS Components Ltd, just a double LED lamp affair, with red and green signal lights and secured it to the top of my machine. Very industrial looking it is too. It was just a bit of fun using the USB I/O to control the lights in my spindle warm up routine, but it does do the job rather well, and I can see exactly when the 5 minutes is up simply looking at the tower.

I plan to use this M64 and M65 code in other projects, so I will be able to see at-a-glance when each part is finished. I discuss this modification in a later chapter.

### 8.3.4 G28 command

Rather like the G30 position we are using as our 'safe space' location, the G28 code is another pre-configurable location, which is stored, supposedly, in non-volatile memory. Unlike the G30 position, you cannot simply set the position using the button on the PathPilot® screen, nor can you manually make the machine 'go there' by using the button on the main screen. But it is certainly a useable location to call from within your program, so with the use of Visual Studio as an editor, you could add some code at the end of a program to move your table and spindle to wherever you wish. However, on my Tormach® mill, I didn't find G28 storage to be non-volatile, and I am not sure why.

### 8.3.5  G53 command

G53 is something I have, so far, not mentioned. In truth, I am not entirely sure there is much benefit in doing so. However, in my dealings with Tormach® when I had an issue with long tools crashing into my workpiece after a tool change (which subsequently ended with an update to PathPilot® to fix this), I came across this command. And it could be considered quite interesting, depending on your point of view.

G53 is the three-dimensional space which is based on the 'absolute' co-ordinates of the machine. Putting it another way, although we might decide to set our own origin and label it G54 work offset, but we can never set G53 workspace because it is fundamental to the machine itself. I believe the machine 'thinks in inches all the time', even if we are in G21 mode, but don't quote me on that.

The machine always knows its limits. Supposing we fixed some work in a vice and probed where it was in G54 workspace. And then we inadvertently gave the machine a command which was to move in the X axis by, for example, 1000mm. This is way beyond the allowable range of movement in X, and the machine would hit its limit switch, so rather than just let us go ahead and test the switch out, the machine already knows where our G54 space is compared to its G53 space and will always let us know if it thinks we are asking the impossible. And sometimes we will get an error message telling us it can't be done, when we press 'Cycle Start'. That's worth being aware of for starters, but there is one more thing about G53 I want to share.

If you type into the MDI line, on PathPilot®'s main screen, the command G53 Z0, the spindle will rapidly traverse upwards until it reaches its upper limit. This position is where Z=0 in the machine's absolute co-ordinates, so the command will always have this outcome, no matter what workspace us humans have set the machine to. Occasionally you might see this command in program code, so now you know what it means. Notice how, in machine terms, all its Z space must be less than zero.

### 8.3.6 G30 command

I already explained earlier, Tormach® have programmed the tool change macro so that if the machine stores a tool, and then retrieves another, the spindle will go to a position which is the higher of either the current tool change height (quite low on the Tormach® 770MX) or the G30 height which you set. That is why it is useful to have a G30 which not only places the work out of the way, to the right, away from the tool changer, but also places the spindle up out of the way so, in case we have retrieved a long tool, then it can happily be sent on its way to the next cutting operation, without risk of crashing into the work. Oddly, the Fusion 360 post processor does not appear to take account of this issue and does not send the tool height to the 'clearance height' you specified in CAM before moving the tool in X and Y towards the work. Hence this Tormach® update is important.

If you merely store a tool, but do not retrieve one, (there is the option to do that with a button on the ATC tab on PathPilot®), then the spindle rises even higher to the G53 Z0 position, as high as it can go. The theory being, likely the next thing you might wish to do is load a tool manually, so this maximises the available space for you to do this.

I hope that helps in the understanding of the differences between work co-ordinate systems we choose, and those inherent to the machine. I finally make the point that when you store a G30 'safe space' position, you are storing it in non-volatile memory which is based on G53 space. The machine knows where you have placed your G30 chosen place, no matter which work co-ordinate system (WCS) you are currently working in, G54 or whatever.

### 8.3.7 G54 command

The nine work offsets from G54 to G59.3 listed above may all be seen in the work offset table on the Offsets tab in PathPilot®. And you can amend the data manually if you wish to. One example of this being useful might be if you were using two offsets for two separate pieces of work held in the same vice, and the Y co-ordinate for both offsets is the vice's fixed rear jaw. Well, you wouldn't need to probe the vice jaw twice, once in each workspace; instead, you could simply type the same number into the next row. There are many times you might wish to change the data in this table.

Although G54 is the first of many workspaces we can define, it is also important to appreciate they may all be used as commands too. By this, I mean, in a program you will often see G54 (or one of the others) and this just commands the machine to think in G54 space for the operations which are about to follow. Alternately, you are quite at liberty to type G54 (or one of the others) directly into the MDI line (manual data input box) and this instructs the machine to think in this workspace straight away. When would you choose to do the latter? Usually just before you probe a piece of STOCK with the passive probe, to place the machine into the workspace in which you require the probing data to be stored.

In Fusion 360 CAM, when we begin a new setup, there is quite a lot of data to enter about the size of the STOCK and where our chosen origin is going to be and, of course, which origin it is going to be labelled. Any one of the 'G54' type work offsets will do. It is worth pointing out that Fusion 360 uses a numbering system starting at 0 (zero) with increasing integers to choose the workspace, and it is a little confusing at first until you realise the first one, namely G54, is identified by 0 or 1, and then G55 becomes 2, G56 is 3, and so on. Once you realise there is this anomaly just for the first one in the list, it is no issue, but it caught me out a couple of times.

8.3.8  G73 up to G89 canned cycles

I briefly mention these just so you know they exist but, almost certainly, you will never need to use them or deal with them. The reason being modern CAM software, such as Fusion 360, does all this for us.

For example, if you wish to drill a hole then, within Fusion 360 CAM setup, you will choose a DRILLING operation, and then there would be lots of data the software asked you to enter. Obviously, where is the hole? how deep do you want it? but it then goes on to find out if you want to drill it in one go, or peck at it, and even whether you wish to 'allow for' the tip of the drill or not. If you allow your mouse to hover over many of the options, there is a comprehensive and diagrammatical explanation of the choices you face. All these chosen parameters go into the melting pot and Fusion 360 comes up with the code. You don't really need to know how it did it. I am sure it is using canned cycles, but you really can't tell. This is the beauty of modern post processors.

### 8.3.9 M1 command

The last code worthy of a mention is M1. On the main screen of PathPilot®, there is a button marked 'M01 BREAK' and it has quite a useful function. Wherever there are M1 (the extra leading zero is not required) commands in the program produced by the post processor, the machine will stop at that point, and it will wait for the operator to press the CYCLE START button to continue the program. This is only the case if the 'M01 BREAK' button is active and has its green light illuminated. If you switch it off, the program will totally ignore any M1 commands it comes across.

This is great when you are prototyping and need to be watching over the beginning of each separate operation, but you trust the machine to behave itself in between those moments. Once you are sure the whole program runs without any danger, you can de-activate the 'M01 BREAK' button, for subsequent cycle runs.

Alternatively, in Fusion 360, you can choose where the post processor will insert M1 breaks; either between tool changes, between operations, or not at all. I tend to keep the box ticked for 'between tools' so that it will always wait for me to come and watch the start of a new tool operation. Once I am happy, if I am making a lot of the same part and I no longer need this safeguard, it is no bother to switch off the 'M01 BREAK' button. But the choice is yours.

### 8.3.10 Summary

In summary, there is a lot more to G-code and its nuances than I have offered. But I think I have picked out the basics for us to get started making parts. And that is the aim. If you want to find out more, there is so much additional information out there. But my advice, to enjoy using this machine to its fullest extent, is concentrate more on the intricacies of Fusion 360. If you can get up to speed with Fusion 360, you will find the Tormach® will run its post processed G-code extremely reliably and accurately and the limit to what you can do is dependent on your mastery of the CAM software, not the machine itself, and certainly not your mastery of G-code.

Early on, when I was designing some parts, 'doing' the CAM for those parts, and then setting up the Tormach® to manufacture them, I was spending about 8 hours of my time sat at my desktop working through Fusion 360 CAM for every hour at the machine. Of course, if I were to go on to make hundreds of the same part, that wouldn't be such a meaningful statistic because you can walk away in the knowledge the machine no longer needs babysitting, but for me making a few parts as I learned to use the machine, you get the point. When it comes down to it, you really do need to learn CAM to enjoy the full benefits of this machine.

# Chapter 9     Connection to Wi-Fi

First up, you do not need to connect the Tormach® to the internet. The machine works perfectly well 'offline', and you can 'feed' its appetite for G-code by transferring it from your computer using a USB memory stick. Much like any desktop 3D printer.

But let's not kid ourselves, it would be much easier to use if it was a connected machine. There are a few discussion points. There are two tasks why having the machine connected to the internet is a good idea; uploading your program code from the computer on which it was created and updating the PathPilot® software of the machine's controller. You can tick a checkbox to always 'check for updates', so you need never be using out of date software.


9.1   Wi-Fi adapter on a USB stick, supplied with the Tormach® 770MX

The tiny USB Wireless Network Adapter which Tormach® supplied me worked straight out of the box and one of the first things the computer asked me to do was update the PathPilot® software. It was incredibly easy, and seamless.

The Tormach® PathPilot® Console has six USB sockets, four at the rear and two on the front above the screen, high up. They get used up quite quickly, though. Here is what I did:

Rear USB sockets

- ATC control
- USB I/O module
- Keyboard
- Mouse

Front USB sockets

- Wi-Fi Wireless Network Adapter
- Spare - (USB memory stick for G-code transfer)

The reason I placed the Wireless network Adapter on the front was it better faced my router and was generally less tucked in behind the steel casing of the controller, so I was hoping for a better signal. It works well.

## 9.2 Mapped network drive

This is really a great idea if you are on Windows. If your CAM software is on a Mac computer, then I do not know anything about those; I can only tell you of my experiences using my Windows desktop.

I am not going into the technical setup because I am no computer expert and there is all the information you need online, through Tormach® technical data sheets and, also, I highly recommend you look at a video John Saunders at NYC CNC has done on this.

The gist of the setup is you give your Tormach® a name, and a drive letter. Mine is called Tormach® and I have made it the T: drive. Once you have mapped this drive into your home network, things get very easy indeed. On your computer with the CAM software, you can now 'see' the T: drive and any G-code you create can be stored in a folder on the drive. Both the computer and the Tormach® PathPilot® computer must be simultaneously powered to keep the folders in sync since the T: drive only really exists on the Tormach®. On the PathPilot® screen there is a tab called File and there you will find the uploaded new G-code which has come from your other computer over the network. What I really like is if you are running a program on the Tormach® and simultaneously making some changes to the same program on the other computer – maybe you noticed you could increase a feed rate for a particular operation or something like that – then if you update that code, it sits in the T: drive waiting for the Tormach® to finish what it is doing. As soon as it is finished, a 'pop-up' message on the PathPilot® screen advises you there has been a change to the file and asks whether you wish to upload this into the working memory. All very tidy.

## 9.3 PathPilot® hub

Tormach® have created a 'virtual' PathPilot® Console online and once you have created a login identity, you can upload your own code to it to simulate using the machine. It is free of charge to use and is an excellent means to explore the PathPilot® system before you have even purchased a real machine.

The link for the hub is here: https://hub.PathPilot.com/about

In truth, the simulator is fine if a little slow as it runs across the internet. Also, the motion of the milling cutter as depicted on the PathPilot® screen is not overly clear on the actual PathPilot® Console, let alone on the simulated version of it. It is fine for giving you an idea of what is happening but no real detail. If you are running Fusion 360 CAM software on your main computer, the simulation within this is remarkably accurate and detailed, and leagues ahead of the Linux simulation, to be honest.

I don't tend to use the PathPilot® Hub because I don't find it so useful. However, it is worth mentioning the Hub is built into the PathPilot® Console on the Tormach® so whatever you upload onto the Hub cloud server, is immediately retrievable on your Tormach® machine, thereby making this another good method of transferring programs to your machine over the Wi-Fi connection. If the mapped network drive method is proving difficult to setup, this is a completely different way to do it. And it is free of charge to have the Hub account.

## 9.4 Dropbox

Many of us have a Dropbox account and it is yet another means by which we can transfer files to the Tormach®. I was very keen on this method since it meant I could store my new code in a suitable folder called 'Tormach® G code' and it would be there immediately I switched on the Tormach® machine. Both machines would not have to be on simultaneously to sync the folder, as Dropbox would be acting as the intermediary.

There is a slight issue though. The Tormach® PathPilot® computer is running Linux, not Windows, so there are certain restrictions in the setting up of your Dropbox account on the Console. Unlike within the Windows environment, you cannot just sync a single folder within your Dropbox account, you must sync the whole account initially, despite what the Tormach® manual says. If you have a huge amount of data, this will swamp the Tormach® drive to breaking point. I had a discussion with one of the engineers at CNC Machine Tools, Snetterton, Norfolk, UK, the supplier, and importer of my machine, and he came up with a fix.

Basically, you need to take all your data out of your Dropbox folder and store it temporarily 'somewhere else'. Then you create TWO folders in your now empty Dropbox account, one called 'Tormach®' and the other called 'Personal' or some such suitable name.

Then go ahead and follow Tormach's instructions and setup Dropbox on the PathPilot® Console and it will sync both folders. They are empty, but they are sync'd. Once this full setup is complete, and only then, you can go into the Dropbox settings on the PathPilot® Console and switch off the sync setting for your folder called 'Personal'. And then, back on your main computer, you can put all your personal data which you saved elsewhere back into the 'Personal' folder where it will no longer sync to the Tormach® machine.

A little long-winded, you end up with this slightly odd 'sub-folder' affair, with all your personal data contained in it. There is also the added issue of where to put the data while doing this, without a certain amount of risk of losing it. It is not a simple problem to solve if you have a large amount of data. I have not done this yet because the mapped network drive method is working perfectly for me. The question was raised about having a second Dropbox account but I don't believe they permit two accounts on the same computer, so this is not a simple solution either.

If you currently don't use Dropbox at all, then I strongly recommend this method for transferring your G-code from main your CAM computer to the PathPilot® controller because I am sure it will work extremely well. Probably the best way in fact, especially because you don't need both your computer and the Tormach® powered up for the sharing of files to still function. It is just not that convenient if you are already a 'heavy' user of your Dropbox account as, indeed, I am.

## Chapter 10    Tool holding

It is quite amazing how much money you can end up spending on tooling, which encompasses cutting tools and tool holders. Although I purchased the 'starter kit' from Tormach®, which included some BT30 collet type tool holders, some 'fixed size' end mill holders and a decent tray which holds twelve BT30 holders, I didn't buy any cutting tools from them as they only appear to supply imperial sizes for their North American market.

The starter kit is quite expensive, but I do use almost everything in it. Surprisingly important is the 'tool tightening fixture', which rigidly bolts to a bench and is invaluable for holding a BT30 tool holder while you tighten a tool in it. Also, the first time you use one, you will need to assemble the 'pull studs' onto each of your BT30 tool holders and even this is difficult if you don't have the fixture to hold it safely.

If you are new to milling, then the starter kit is a good way to educate yourself about the various ways of holding milling cutters. If you have previous experience of milling with manual machinery, then you will already be aware of some of the methods, many of which no longer apply in the CNC milling world. So, things like R8 collets you find on Bridgeport machines and, later, machines imported by the likes of 'WARCO', and 'Chester', are no use to us for this project. The tooling collection you may already have built up, generally will not be transferable to the CNC machine. The earlier Tormach® TTS tool holder design used the R8 collet system, and I understand the smaller 440 mill by Tormach® still uses this, but for this project, we are considering the BT30 spindle only.

Personally, I have a very small 'WARCO' mill, the spindle of which has a 2MT Morse taper. Fitted in that taper, with a drawbar, I use a 'Pozilock' type collet chuck for my milling cutters, sometimes known as a 'Clarkson' chuck. Cutters for this must always have a threaded shank – some sort of 20 TPI Whitworth thread. Again, this is fine for the manual mill, but of little use for our shiny new Tormach®. (Interestingly, the threads on the shank of the Clarkson type milling cutters do not protrude proud of their smooth shank diameter so these cutters could, in theory, still be used in an ER collet chuck, more of which later. But my advice is to buy some fresh new tools.) This Tormach® is fiendishly accurate, and we want to get the absolute best results we can. We will only manage that with high quality sharp tools, which are exceptionally accurate in their ground diameter.

A reminder the Tormach® 770MX CNC mill is only offered with a BT30 spindle. This means there are only a few tool-holder options we can buy which will fit the spindle and I will take you through the most useful ones I have come across, and indeed purchased.

## 10.1  The basics of the BT30 spindle

The BT30 tool holder fits into the spindle inside a corresponding taper which is non-locking. In other words, the taper angle is such that the tool holder will never get stuck in it. Compare that with a much finer angle Morse taper, as an example, which dictates the need for a means of physically releasing the tool holder because it naturally gets stuck.

You must separately purchase 'pull studs' which screw into the top of the BT30 tool holder, and it is these which are grabbed by the drawbar mechanism. In the case of the Tormach® 770MX machine, this is pneumatically operated. Compressed air at about 90psi forces the mechanism to release the grip on the pull stud, and mechanical spring pressure clenches it back again when the air is vented. The spring return pressure is made up of a stack of conical spring washers known as Belleville washers and is adjusted very precisely so the tool holder is held securely but also released on cue without fuss. In this way, automatic tool change can occur smoothly and reliably.

The BT30 spindle has cut outs to accept the drive lugs machined into the tool holders. I don't know, but I suspect the drive is due to the friction from the taper, not the lugs, and the lugs are more likely there to prevent any turning rather than taking up the drive. Nevertheless, they must engage properly. There is a knack to manually inserting one of these tool holders into the mill. It can't be fully inserted until the air pressure is on, and then it will not fully seat until the air pressure if off again, and the clasping sensation of the drawbar can be felt and witnessed.

Note for those machines which have no automatic tool changer, the button to release the drawbar is a mechanical pneumatic switch on the side of the head. Whereas, if you have installed the automatic tool changer you will already know you had to make some radical modifications to your machine, chuck that pneumatic switch out, and re-plumb the piston assembly through to the pneumatic solenoids within the automatic tool changer electrical cabinet. The equivalent manual release button then becomes a small electric switch on the side of the automatic tool changer cabinet. It works well and is positioned ideally. The only difference this causes is, even if you have full air pressure connected to the machine, you are completely unable to release the drawbar until you have fully powered up the PathPilot® Console and reset the Tormach® machine. This is because the automatic tool changer needs to communicate with the controller before it will function. You cannot make any manual tool changes until the machine is in a fully reset state.

**BT30 tool holder with ER20 collet chuck**
Copyright© photograph courtesy of Tormach® Inc.

In the example you can see the tapered shank of the tool holder. The top half always looks the same. The bottom half is where there are differing options, depending on which type of tool you need to fit. The one shown is an ER type collet chuck, which I shall come onto shortly.

You can see the pull stud has not yet been fitted. These are separate to the tool holder itself because they can suffer from wear and tear as they repeatedly engage with the drawbar mechanism. So be sure to lubricate them a little and keep an eye on their condition, and maybe carry a couple of spares in the drawer.

I have found an excellent source of pull studs from a supplier in the name of 'LANTRO JS' which sells on Amazon in the UK. Quality is high and the price is excellent.

**BT30 pull stud from LANTRO JS**
Copyright© photograph courtesy of Lantro JS

When you choose BT30 tool holders, there is a parameter called the gauge length which is quite important, more so, since the Tormach® 770MX does not have a huge amount of working space in the Z direction. Once we fit a tool holder into the spindle, we immediately begin to 'use up' that space. And the longer the tool we are using, again the less working space we have remaining. It is quite important, then, to procure short stubby tool holders if we need to make parts which stand tall on the table, or in the vice. Otherwise, we might run out of space.

Also, since the automatic tool changer carousel is quite low on the 770MX machine, the shorter the tool holders the less they will hang down from their position in the carousel, and the less potential there is for tools to crash into any work.

Thirdly, it makes sense if you are side cutting using an end mill, (thus exerting sideways force on the spindle bearings), the shorter the length of the combined tool and tool holder, the less leverage is acting on the spindle, so that ought to be advantageous, both in terms of spindle bearing wear, and in terms of tool run-out due to bending forces.

So, what is this gauge length? It is quite difficult to measure on a BT30 tool holder since the flat area just beneath the taper does not sit flush against the spindle head. There is a small gap because the taper is the limiting fit. Roughly speaking, the gauge length of the tool holder is the distance from the widest part of the taper all the way to the bottom end, where the tool fits. Approximately, if we placed the tool holder into the spindle without any cutting tool inside it, then the bit we can still see is the gauge length.

We will find that 50mm or 60mm is about as small as they get, but that is quite good. My advice is always buying tool holders with this sort of gauge length if you are able.

## 10.2 ER collets

There are different 'family' sizes of ER collet chucks and the example illustrated is an ER20, which is most suitable for the Tormach® 770MX, I would suggest.

It is worth separately researching the ER collet system – there is plenty of excellent information online. Get to know a bit about how they work and why they are so good at what they do. This is especially useful because they are also excellent work holding devices, as we shall come to see a little bit later.

There are some salient points, very much worthy of note. ER Collets are rather a clever design in that they can clamp down from their 'labelled' size to a diameter nearly as small as the next smaller size in the range, whilst keeping a parallel hold on the tool being held. Compare this to R8 collets or the 5C range of collets, which absolutely are unable to do this, and you will appreciate just how good ER collets are.

**ER collets**

The collets are manufactured in such a way that each neighbouring slit starts from the opposite end, so when the collet is tightened down and pushed into a taper, the collet can shrink in diameter from each end in equal measure, thereby keeping the parallelism. By toolroom engineering standards, the ER collet design is relatively young, around forty years.

An important thing to note is the way in which the design of the nut which tightens the collet in the chuck is slightly at odds with how you'd expect. Indeed, if you look inside it, you will see a lip which is not centrally placed. You must engage the nut with the collet BEFORE you place the collet into the chuck. It will usually be a bit of a fiddle and sometimes may 'click' into place. If you make the mistake of placing the collet into the chuck, and then tightening the nut over the top, you will still tighten the collet up, by pushing it further into the tapered hollow, but you will have a real problem getting the collet out again. The idea is the nut is pre-assembled to the collet, and when you undo the nut from tight, it pulls the collet out. It is counter-intuitive but vital.

The ER20 size is an excellent starting point for the Tormach® 770MX. The number represents the diameter of the tapered hole in the chuck, so for ER20 collets, the taper at its widest point is 20mm.

ER20 collets are made for every 1mm increment from 1mm diameter up to 13mm, so a useful range. You probably won't want to be using end mills much above 10mm or 12mm anyway, since you are limited by the relatively low power of the spindle motor, so there is little to gain from being overly ambitious.

The Tormach® starter kit contains six of their BT30 ER20 collet chucks, which are 60mm gauge length, with a standard nut. A spanner is supplied. You can't have too many of these, and I have since gone on to buy six more. They are a perfect fit for much of what I wish to do.

I have bought a few other brand BT30 ER20 collet chucks, and some have a castle type nut requiring a special spanner, so be aware of this – it is quite common in ER collets to have various types of nuts, but they all submit to some sort of 'standard'.

The magical thing about ER collets is they can hold just about anything that is predominantly round. This means any end mill will work in them, whether it is one with a purely round shank, has a Clarkson thread at the end (the thread would be ignored in this case), or even the sort which has a notch in its side. If there is enough round shank to grip, the ER collet will be ideal. You can also use them for drills. I much prefer to use a collet chuck for a drill instead of a drill chuck since I believe it holds the drill tighter, straighter, and with less run-out.

It is important to place any tool into a collet, such that the entire length of the collet is clamping on the diameter. You don't want to be just biting at one end since the collet would struggle to stay parallel, and the tool would wobble free as soon as you tried to use it.

One final point I wish to stress is you will find yourself buying many end mills which are standard sizes, such as 6mm and 8mm diameter. And sometimes, very small diameter drills and end mills have standard 3mm or 4mm diameter shanks. For this reason, do not be surprised when you realise, you'll need a few more collets in those sizes and you will need to order individual collets rather than sets. Collets come in different levels of tolerance for run-out, so the price can vary quite a lot, 10 micron run-out being standard kit, 5 micron run-out for high performance, and 2 micron run-out versions for extreme precision industrial needs on a different scale.

The ER20 collet system is going to be the most useful means of holding your cutting tools. But there are other tool holding options we can buy which are based around the BT30 holder.

## 10.3   Basic drill chuck

Tormach® supply two BT30 drill chucks with their starter kit. The chucks can hold up to 8mm diameter drills, are 'keyless' but do need tightening properly using the special 'hinged peg' spanner which engages with slots in the side of the chuck. The gauge length is 80mm, slightly longer than the ER20 collet chucks.

These are fine quality and work well for smaller drill sizes but, if I have a job with any drill size which is very near to a collet size, I will always prefer to use the ER20 collet system over the drill chuck. The drill would be more tightly held with better concentricity, I believe.

Drill chucks are only for drills. NEVER put an end mill or indeed any milling cutter into a drill chuck – they are not designed to take any sideways (lateral) force.

## 10.4   End mill holders

End mill holders are the sort which are designed to take those end mills which have the notch cut out of the shank, or sometimes just a ground flat. I think of them as 'fixed size' end mill holders because the hole which takes the end mill is an exact sliding fit for standard sizes only. The notch (or flat) lines up with a grub screw which tightens the cutter in place so it cannot turn. Also, it cannot be pulled out. Bear in mind the cutting forces will always try to pull any cutting tool from the holder. This is called the Weldon system, and cutters suitable for these sorts of holders are called Weldon shank cutters. The central hole in end mill holders has a diameter and concentricity of high accuracy making these a sturdy option. A cutter which has no Weldon flat is unsuitable for these, however, since the grub screw would be unable to stop the shank from turning or being pulled out.

End mill holders are usually quite compact, since there is very little mechanism (just a grub screw) so the gauge length is often smaller. The Tormach® offerings have gauge length of 50mm which might be useful in some situations where space is limited in the Z axis.

The starter kit provides four of this type of BT30 holder, two of one size and two of another. Since there is still some confusion from Tormach® in the USA about the provision of a metric version of their starter kit, I cannot say which sizes these will be – I negotiated with the UK importer my own choices. To be honest, I use these less frequently because I find most cutting tools I buy have smooth shanks requiring the use of an ER collet.

10.5  Face milling cutters

This is where BT30 tool holding gets interesting. There appear to be a few different standard ways of holding unusual tools.

I wanted a face milling cutter which was quite wide, used indexable carbide inserts, which are replaceable, to experiment with how good a finish I could get on various metals. I chose a 40mm cutting diameter milling body with five inserts by doing some searching around online (different to that shown).

**Example Face milling body, with seven indexable carbide inserts**
Copyright© photograph courtesy of Cutwel Ltd

The milling body comes as a separate entity which needs an additional part, referred to as an arbor, to fit into the machine spindle.

The tool supplier will stock different arbors to accommodate different spindle sizes, so in our case, this would be BT30.

You must choose your indexable carbide inserts separately since there are many different types for cutting different materials. It quickly becomes a complicated choice. There is even guidance for the correct torque you should use to tighten the Torx screws securing each insert to the body. This is industry level stuff and I figured 'sensibly tight' would be OK. It was.

One of the interesting aspects of these face mills with the indexable inserts is you don't have to fill all the locations. In the extreme, you could just fit one. In this configuration, this is just the same as an old-fashioned 'fly cutter' so with a very slow feed rate, you ought to be able to get a fantastic finish. The inserts seem quite expensive until it dawns on you each one has two lives as they are reversible. Later when I discuss feeds and speeds calculations, you will appreciate how varying the number of indexable inserts is accounted for in the mathematics, and thus how fast you can use the tool.

**BT30 Face mill arbor to suit the milling body above**
Copyright© photograph courtesy of Cutwel Ltd

The face milling body fits onto the arbor and is retained by a hex-head machine screw inside the small flange you can see. The projecting lug fits into a slot in the milling body, so it cannot slip. This arbor is called a 'tenon-drive' arbor for this reason. Once again, the BT30 pull-stud is not shown fitted in the photograph.

## 10.6 Collet chuck for taps

The Tormach® 770MX is quite clever and can perform 'rigid tapping', sometimes described as 'synchro tapping'. Fusion 360 can work out all the CAM code for us, so it appears simple when in fact the machine is doing something quite wonderful and complex. The spindle drives the tap into a pre-drilled hole and the Z axis is driven downwards in synchronism with this, to suit the pitch of the thread being cut. The clever part comes when it gets to the bottom of the required thread and the spindle rapidly reverses its rotational direction and at the same time, again in exact sympathy with the thread, the Z axis drives back out of the hole leaving a perfectly tapped thread.

To help with this process, it is best to use special ER collets. These look identical to normal collets at first glance, but the central hole has a squared-out portion at the rear which accepts the square-shaped shank you always get at the top of a tap. This makes it impossible for the tap to turn, so the collet cannot 'lose grip' of it, vital for accurate thread forming.

Secondly, there will always be some tiny error the split second the spindle has to reverse out of the hole, so you can purchase special tapping chucks which have some built-in rubber bushing which offers a little bit of 'give' in the rotary sense, just enough to iron out any shock to the tap when the rapid direction change occurs. This extends tap life and improves the thread form. These tap chucks are called 'synchro' tap chucks, and they are quite expensive.

I bought one which can take collets to suit taps from 3.5mm to 8mm metric thread sizes, which should be enough of a range to keep me occupied for some time to come. It uses ER16 collets, slightly small in diameter to my ER20 collets, and I had to purchase a set of collets to suit the range of taps I would be using. When you are purchasing taps, the size of the 'square' on its shank is an extra factor to ensure you have the correct collet size and is often referred to as the 'K' dimension.

**BT30 'synchro' tapping chuck, with ER16 tap collets**
Copyright© photograph courtesy of Cutwel Ltd.

The rubber bushing inside the synchro chuck is presumably behind the red band. You can't feel any movement at all; it is to all intents and purposes a rigid fixture, but I appreciate they have calculated exactly what is needed. It works extremely well, and I have tapped some mild steel with it using an M8 machine tap at 200 rpm which I regard as impressively fast for tapping. It's a valuable addition to my toolkit. Having proven how good it is, I have since gone on to procure a range of machine taps of varying sizes together with appropriate ER16 tap collets.

# Chapter 11  Cutting tools

It is highly likely you will already know an awful lot about milling cutters; after all you are about to, or are considering, getting into CNC milling so it would be surprising if you had no previous experience of a manual mill. But the subject of modern CNC tooling is certainly an evolving technology, and I think there are some aspects which are worthy of a quick refresh, especially if you are like me and you don't work in this sector.

## 11.1  End mills, not slot mills!

If you've been around a while, things have changed quite a lot over the last few decades, and I trust you'll forgive me for a quick round-up of the confusion I was personally experiencing about end mills. It's quite possible I am not alone in this. If you are young and eager, it's possible you may still not appreciate all the nuances on this subject; either way, I hope the following discussion is of interest.

I have a very small manual mill at home, which I bought from WARCO with a set of 'high speed steel' (HSS) end mills which have served me well for years. Except, I was always under the impression some of my cutters were called 'slot mills' and some were called 'end mills.' My understanding was a slot mill could plunge vertically into material, but an end mill could not. The reason for this is the slot mills were usually 2-flute cutters with one of the flutes' cutting edge 'over centre' so it could still cut whilst plunging. In my collection I have some end mills which happen to be 4-flute and their bottom-centre has no cutting edge at all. If you tried to drive them vertically downward in a plunging action, they would just graunch into the material and break.

But when you begin to 'surf the internet' and look at the latest cutting tool suppliers' online offerings it dawns on you the expression 'slot mill' doesn't really mean what you thought it did anymore. Modern techniques for manufacturing cutters mean it is quite easy to produce 2-flute, 3-flute, 4-flute, even 5-flute cutters which can have one of the flutes 'over centre' at their base so they will easily plunge cut. And they refer to all of these as end mills.

It turns out the expression 'slot mill' is now reserved for what I would call a Tee-slot cutter. If you look at Fusion 360 CAM software and click on tool library, then click to add a new tool to your library, you will be presented with a choice of cutter types, all with very clear illustrations.

- Ball end mill
- Bull-nose end mill
- Flat end mill
- Face mill
- Tapered mill
- Radius mill
- Engrave / chamfer mill
- Dovetail mill
- Lollipop mill
- Slot mill
- Thread mill

The first three on the list are all end mills. I want to expand on the differences. Both the ball end mill and the bull-nosed end mill have a radius at their end meaning where their fluted sides reach the bottom, there is no sharp corner, whereas a flat end mill is ground to a very sharp point at the end of each flute and will cut into a corner almost completely. The ball end and bull-nose end types will leave a radius in a corner which cannot be removed, which is sometimes not a problem, and other times quite aesthetic, so oftentimes these are desirable.

The difference between the two is simply with the ball end mill, the radius of the tip of the tool is the same radius as the tool itself, so the end of the mill is a complete half-circle, a half-sphere in fact when the tool is spinning of course. Put another way, if you plunged this tool into material by an amount equal to the radius, and then proceeded to cut a horizontal slot, you would produce a slot which was the tool's diameter wide with a cross section a perfect half-circle.

A bull-nosed end mill, on the other hand, still has the end of the tool radiused but only slightly. For example, you might have a 6mm diameter bull-nosed end mill with a 0.5mm radius. If you cut the same slot as before, the cross section would be a 3mm deep slot, 6mm wide, but without sharp internal corners, as there would be a very subtle 0.5mm internal radius. That might be fine, and it would certainly look nice, but also consider the additional benefit of using tools without sharp points at the ends of their flutes; the cutting stresses are dispersed away from the tips which makes them inherently stronger. There is less chance of tool tip breakage with these sorts of tool.

Unless you have a particular need for a very sharp internal corner cut, then it is worth considering bull-nosed cutters for a lot of your milling. If nothing else, it will not surprise you to also learn you will cut yourself less frequently with these rounded edge tools, than with traditional end mills, something to which I can attest.

I mentioned the number of flutes earlier. Traditionally, I was always taught 2 flutes for softer materials such as Aluminium and 3 or 4 flutes for steels. With a 2-flute cutter, there is more room for chip evacuation, but with more flutes, there is greater core strength. To be honest, I think I am way out of date on this. Tool technology has moved on with such speed, the science is amazing. It is worth researching on the tool suppliers' websites for all the information you might need. With all the modern cutting tools for sale, there are now very clear spreadsheets from the manufacturers with 'feeds and speeds' data for each tool, and for each type of material you wish to cut. This is not hype. They have done the research for us. If you program your CAM operations according to their guidelines, you will get the result you want. It is explicit these days what you should do. So, going back to my question of 'how many flutes is best?' it is more likely better you search through the data, and you will find tool options to suit your job with varying numbers of flutes, flute shapes, helix angles, materials, coatings, and all manner of extremely technical advances.

There is a choice to be made on what the end mills are made from, high speed steel (HSS) or tungsten carbide which seems just to be called carbide nowadays. I was keen to try carbide tools as I had no previous experience of them and I purchased a selection of 3mm, 4mm, 6mm, and 8mm end mills, flat ended, bull-nosed and ball ended and have had great success. Carbide tooling is high-quality, tough, and wear resistant but I understand it is prone to failure from temperature shock, so care is needed to keep flood coolant on target. At the hobbyist level, I shouldn't think there is much between HSS or carbide tooling but, if you are manufacturing parts in greater quantities, I expect carbide tooling will last longer.

## 11.2 Chamfer milling cutter / spot drill

Fusion 360 tool library makes the distinction between what they term an 'engraving / chamfering' tool and what they term a 'spot drill'. In truth, they are similar, and I have found it difficult to understand the subtle nuances there must surely be. Having said that, I have bought an excellent cutting tool which I use for both chamfering and spot drilling, and I have told my CAM software to consider it as a 'spot drill' in the library. I will come back to engraving in a later section.

**A 90º carbide spot drill**
Copyright© photograph courtesy of Protool Ltd.

This tool has a 90-degree angle which is probably the most useful angle to buy. I use it for spot drilling, prior to drilling holes, particularly small diameter holes, because I think it gives the drill more persuasion to remain central as it starts the hole. I simply spot the planned hole with about 0.3mm depth, nothing more, just to give the small drill a good starting point. It seems to work quite well.

Of course, this will also countersink any hole if that was required. Additionally, that would be a useful thing to do for a hole that you wished to thread tap, giving the tap an easier start.

But what this tool is most useful for is chamfering and this is one of the joys of CNC milling and something which was much more difficult on a manual machine. The beauty of this tool is it will chamfer any straight edge, curved edge, external corner, even internal corner (with a bit of thought about tool radial offset), beautifully and easily. Researching the internet, I found some useful advice for chamfering using Fusion 360 CAM software. There is a 'chamfer' option under the 2D operations tab, but the suggestion is to use the '2D contour' option instead. Within the edit box for this operation, you will find a chamfer tick-box within the 'passes' tab and this works well. You don't even need to put a chamfer into your design of the part; it can be a sharp edge in the design, but the CAM process is still able to pick out this edge and chamfer it nicely.

I mentioned tool radial offset (they call it 'tip offset' in Fusion 360). I have a 6mm diameter spot drill with a 90-degree tip. There are two flutes. This means each cutting edge is just over 4mm long (a bit of Pythagoras, nothing more) so between different jobs I attempt to vary the offset of the tool within the CAM software, thus attempting to utilise different parts of the cutting edge and avoiding wearing the same spot all the time. The software will do the maths for me and still chamfer the correct amount off the edge of the part. An added complication arises when you are trying to chamfer an 'inside' corner. The more radial offset you specify, the wider the diameter of the tool at the point the cutting edge meets the chamfer. For an inside corner, you need to be sure this diameter is smaller than the actual corner you are intending to chamfer, otherwise it will overcut in the corner and look very wrong. It's an obvious point, but one worth thinking through, and once you look at the Fusion 360 software, it will all become clear and solvable. The Fusion 360 slow-motion simulation comes into its own for checking this, and for improving your comprehension of what each parameter change does.

One final point on spot drills which I learned the hard way. The first one I purchased didn't look exactly like in the photo; it didn't have the sharp point. In fact, it was a cheaper model which was slightly rounded off. I didn't appreciate the importance of this at the time. I told the tool library I had a 90-degree spot drill and measured the tool offset using my ETS probe. The data auto-populated into my tool library within PathPilot® so the machine now knew the tool's length. Except it didn't. I started using the tool to chamfer edges and the chamfers were huge, much more than I thought I had specified in my CAM program. A millimetre or so larger. What had I done wrong? I am sure you are already ahead of me. The software can only think of the spot drill as a perfect 90-degree sharp point so, when I used my probe to measure its length, the machine now considered that was where the perfect 90-degree point started. And of course, this was wrong.

For an immediate fix, I simply estimated how much rounding off the tip there was (about a millimetre), went to the PathPilot® tool library, read the length value in the table, added 1mm, and typed the new value into the table. Later I purchased a nicer quality spot drill which has a sharp point and probes accurately into the tool library. This allows me to get into tight corners with a small chamfer if I need to.

## 11.3 Face milling

It is possible to face off the top surface of a part using an end mill and using new, sharp, carbide end mills I have managed to get some extraordinarily rewarding results. Of course, you are relying on the Z axis of the machine being exactly perpendicular to the table to get the perfect result and my Tormach® 770MX is 'good enough for Government work' as the saying goes.

But looking at all the exciting options in the online catalogues, it seemed rude not to at least try one of the face mills which use indexable inserts and, as already mentioned, I purchased a 40mm diameter face mill with room for five indexable cutters.

**An APMT type indexable insert**
Copyright© photograph courtesy of Cutwel Ltd

There is a whole science to the many different types of indexable inserts. They are moulded with built-in rake angles to suit each type of material to cut and have different shapes for different cutting operations. The options are vast, all backed up online by feeds and speeds data for every single type of insert. I bought some APMT type inserts, and specifically chose some to cut Aluminium alloy.

The beauty of getting a wide diameter tool is I can face off the entire top of a part in one pass rather than having to go back and forth and leave witness marks. Within Fusion 360 CAM software, the face operation is the most straightforward to set up. And it gives the option of doing a sizing cut (or cuts) followed by a final finishing cut and this all works wonderfully.

The other advantage of these indexable systems is the metal tool will last forever and it is only the inserts which will need replacing at some stage. As I mentioned previously, some people remove all but one of the inserts to effectively convert the tool into an old-fashioned fly-cutter, the thinking being not all the inserts can be in perfect alignment so this should give an even better finish. I tried this but couldn't find a noticeable difference. I think they are aligned incredibly accurately, so I wouldn't worry too much about that.

## 11.4 Engraving

There is a useful section within the Conversational Programming part of PathPilot® all about Engraving. This gives you the opportunity to just engrave some letters onto any piece of material in your vice, at short notice, without having to spend time designing the work in CAM. I still have a lump of scrap Aluminium sitting next to the Tormach® keyboard, the ends still rough-sawn unfinished, with just the top faced off and the word 'TORMACH®' engraved in it. It was literally the first thing I did on the machine. I can't throw that piece of metal into the bin. It's part of my little machine's history.

I never used that part of Conversational Programming again, strangely enough. I didn't find it so useful after all. But I am making lots of parts which I engrave. The Tormach® 770MX is excellent at it, but I find the Fusion 360 CAM software is more powerful in its control of what, how and where you engrave. I will come back to this because there is much more to this aspect than you might first imagine. But let's discuss the cutting tools you might choose for engraving first.

### 11.4.1 Engraving tool types

When you investigate CNC engraving, there appear to be at least four different styles of engraving tool which will work on a CNC machine. Some of the tool suppliers do not offer all the options and have their own preferences, so be sure to look around to get the effect you are after. These are the four main sorts I found:

- Spring loaded 'drag' tool, with the spindle not rotating
- Very sharp-pointed tool, with very fast spindle rotation

- Small diameter ball nosed end mill, with spindle rotating
- Spot drill or centre drill

I have not tried the spring loaded non-rotating sort and know nothing about it apart from what I have gleaned from YouTube. It seems to be scratching the work with a sharp tool, with the spring tension adjusted to suit how deep the engraved gouge goes. I might try it one day, but I need to add it to my list of things to test.

I have tried both the sharp rotating tool and the ball nosed end mill rotating tool and they give very different results. There is a very interesting YouTube presentation by John Saunders of NYC CNC on this, which I highly commend.

Basically, the sharp tool is like a 'D' cutter, a pointed cone ground away on one side to reveal a sharp 'half a cone' which is driven around the material with the tip just beneath the surface, as you'd expect. The consequence of this is some material is thrown upwards creating a non-smooth surface afterwards. I guess it has something to do with Poisson's ratio, or something like that. Rather than cutting material away, you are engraving so you should expect to get some 'raised' burr. The question arises; is this desirable or not? For a part, probably not I would venture. One solution might be to leave the faced surface slightly over-size, engrave, then re-face the surface again to the correct dimension, which should smooth off the edges, still allowing the engraving to be seen. That doesn't sound particularly efficient though.

I then went on to try some engraving using a small diameter ball nose end mill. In this case, you certainly are milling material away, rather than engraving, so you end up with a smooth final surface which does not rise above the previously flat surface you had prior to the engraving process. Being pedantic, then, this probably is not truly engraving, but it does give a pleasant finish. Using a 2mm diameter ball nose end mill and setting the depth of the tool to just pierce the top of the faced surface by 0.05mm (a couple of thou if you are working in North America), I have created some very fine designs which are beautifully smooth.

I think the use of a spot drill with a sharp point, or even a centre drill will give good results but, again, it really is a matter of personal taste and judgement so I could write whatever I like about it, you would still need to see for yourself. I strongly advise you to buy different styles of engraving tools and experiment. It is a personal decision as to what you will prefer. Whichever tool type you choose, we now need to figure out how to design the CAM processes to engrave things on our parts.

### 11.4.2 Fusion 360 CAM engraving techniques

I want to discuss how Fusion 360 CAM design fits in here. Before we can engrave anything, we need to place a contour into our part's design, which we can highlight to the CAM operation, saying 'engrave this bit please'. That contour could be a simple line, or some text, or an imported design of our own. Let's discuss some 'text' engraving first.

We need to go back into the DESIGN space of Fusion 360 and add the text onto the surface we would like to engrave. That's straightforward; you start a new sketch, and click Add Text, and a new palette opens with options for text size, font, and positioning. You can even add text around a curved line – simply 'right click' on any line in a sketch and there is an option for 'text on path'. You can spend hours playing and investigating this, and every change you make, you can run the simulator to see what the effect is.

When you click on 'font' to choose one, there are quite a few of the well-known Microsoft fonts and these are represented as they would be in a Word document. These are known as True Type Fonts (TTF). It is important to realise these fonts have a richness of quality which affords them their recognisability. Because of this, they are not simply infinitely thin theoretical lines, but each letter has some thickness to it. If we save some text on the surface of the designed part, using a TTF font, then this becomes the 'contour' in the MANUFACTURING (CAM) environment which we will use to engrave by and, depending on which operation we choose, we will get different results. This is subtle, but worth the time and effort to comprehend, so stay with me.

In the CAM environment, under the 2D tab, there are two options which look promising. Namely TRACE and ENGRAVE. These are different functions. The trace function looks at the EDGE of any contour you highlight to it. Consequently, if you have used a TTF font to produce some letters, the trace function will attempt to engrave around the edge of each letter, leaving each letter with a hollow unengraved centre. For large size letters and using a ball nose end mill much smaller than the letters you wish to engrave, this can be a pleasing effect.

The engrave function is different. It will attempt to guide the engraving tool equidistant between the two lines that make up the sides of each 'limb' of a letter or character. Also, and more importantly, it will try to lower the tool into the material until it thinks the tool has reached those sides within the character. For complex fonts in which the letters do not have equal thickness of lines (an example might be a lower case 't' which is broader in the vertical than the little 'flick' at the bottom), this gets complicated, of course, but all the mathematics is done for us, and the effect is more genuinely one of being 'engraved'. Obviously, this function demands a tool which is a sharp point. If you tried to use a ball-end mill, the software would decline the option, since it could not make the necessary calculations. It is a very clever function, but it does cause the surface to be raised, most definitely not smooth.

I tend to use the TRACE function because it gives results which I can better forecast. But what if I don't want letters which are engraved around their edge and 'hollow'. There is a solution, but it is slightly limited in options. Go back to when you were choosing a font, in the DESIGN space, there were some fonts at the top which were not TTF, but instead were called SHX fonts. This is a special class of fonts, invented by Autodesk I think, which have no thickness at all. They are mathematically infinitesimally thin fonts, a single line. There aren't many of them, since you cannot create so much style without thickness, but they can be useful in this case.

Choose an SHX font, draw some letters on your design, then go back into CAM space and choose the TRACE option again. As we already found out, the trace option follows the line exactly. In this case, each letter is just a line, so the tool engraves the letters with zero width. If you are using a small diameter ball nose end mill, you will get amazing results with this method again, but this time, your letters will not be 'hollow'. And because you were using a milling cutter, the result is a perfectly smooth finish.

One special point to make about using the TRACE operation in Fusion 360 CAM. You must set an axial offset of how far down you wish the tool to pierce into the top surface. I usually choose something like MINUS 0.08mm and the minus sign is vital. If you forget to set an axial offset, then the tool will follow the highlighted contour (the text you placed in the design space) exactly at the height of the surface, leaving no visible mark. You must lower the tool to the chosen amount BELOW this contour.

What about if you want to engrave a logo or a design you found on the internet instead of just letters? There is a bit of a trick to it. Start off with a black and white image of your logo. It is most probably a JPG picture file. That's fine, but we need to convert it before we can import it into Fusion 360 DESIGN space. Fusion 360 will allow the importation of files called SVG files, but not JPG ones. I found a very useful converter website which does this for free.

URL: https://convertio.co/jpg-svg/

I have used it a few times with perfect results. Goodness knows why it is free, but there we are.

In the DESIGN space of Fusion 360, click on the INSERT tab and 'insert SVG file'. Amazing things suddenly spring into life and you end up able to place your SVG image onto your part, wherever you want it, at whatever size you wish. Save this and then pop back into the CAM space, click on the TRACE operation again and you can now select this new imported image as a new contour, around which you can once again engrave. Notice, I chose the trace option again, so if the SVG image had 'width' which it almost certainly will, the tracing function will follow the edge of the image and you will end up with a milled outline which follows the perimeter of your logo.

Engraving text and personal designs onto your parts with the Tormach® 770MX is exceptionally effective and limitless in options. There is so much to investigate. I hope I have whetted your appetite to find out more using your own machine.

## 11.5  Machine taps

I am sure we all have some hand taps in the bench drawer, with straight flutes, probably all the normal metric sizes. In my previous manual machine world, if I needed to put a thread into a part, whether on the lathe or the mill, I would introduce the tap into the hole from a drill chuck in the mill, or the tailstock of the lathe, and gently wind the tap into the work by hand until I knew it was going to be perpendicular 'enough'. Then I would take over by hand with a tap wrench. As you know, you must 'back off' every so often to break the swarf and clear it a little, or else the tap would likely snap when it jammed.

Now I know there are some clever auto-reversing tapping devices invented for the older machines, but I didn't ever experience those, so forgive my exuberance when I think about how the Tormach® taps threads. On the CNC machine, I chose not to use straight fluted taps. They just cannot clear the swarf properly if you are making threads in blind holes. I chose some machine taps which have helical flutes. They are a super clever design because as they screw into a hole to cut a thread, the helical design of the flutes causes the swarf to be forced upwards and out of the hole, thereby removing the possibility of a jam. They work for any through hole too, of course.

**Helical flute machine tap**
Copyright© photograph courtesy of Cutwel Ltd

I bought a set of these for a few of the metric sizes I thought I might need, and they are general purpose for cutting into steel or aluminium alloys, so quite useful.

Watching the Tormach® cut threads by the rigid tapping method at 200 rpm, in and out in the blink of an eye is impressive and I still have a wry smile on my face every time I watch. And the quality of the threads is extremely high.

## 11.6 Thread milling tools

For simple threaded holes in parts, I shall certainly be using my new rigid taping tool and shiny new machine taps. But I have also purchased a thread milling tool to learn about how this works. I am being totally honest with you; I have not got to the stage of cutting threads this way yet.

I wanted to get the book out to you, because I am so excited at how good this Tormach® 770MX is, I just want to spread the word and motivate others to import some more of these formidable machines into the UK and Europe.

But my next big mission is to teach myself how to cut internal and external threads on parts by thread milling rather than using taps. This process promises a lot. The thought of being able to do things like cut a 1.5mm pitch thread around a boss which is 40mm in diameter is mind boggling. I can hardly wait to see it in action.

# Chapter 12   Feeds and Speeds

A background in 'hobbyist level' manual milling (and manual turning) does not truly give you an awareness of accurate feeds and speeds. Experience teaches you roughly what rpm to set the spindle for a cut, but if you feed the work into the cutter too quickly, the machine's displeasure will be offered as instant feedback, with a loud squeal or uncomfortable vibration. An experienced operator of a manual machine has a natural 'feel' for what is right.

The same cannot be said for the CAM software programmer. He or she can hardly have any 'feel' for the machine. The 'feeds and speeds' must be entered into the program and the machine will do what it's told. So those numbers had better be right first time. Or at least close.

## 12.1   The PMK chart

Typically, for a manufacturer's range of milling cutters they will publish cutting data for the variety of materials they are designed to cut. These will be grouped in a standard 'ISO materials' chart colloquially known as the 'PMK' chart because the first three are always those letters in that order.

| Material Code | Metal Type |
|---|---|
| P | Steel |
| M | Stainless Steel |
| K | Cast Iron |
| N | Non-ferrous |
| S | Super-alloys |
| H | Hardened Steels |

**The standard letters and colours in the 'PMK' chart**

For some reason it always starts with P, which is steel. I have no idea who chose the letters.

| | |
|---|---|
| P | Steel, including mild steels |
| M | Stainless steel |
| K | Cast iron |
| N | Non-ferrous including Aluminium alloys |
| S | Titanium and other tough materials |
| H | Hardened steels |

The information contained in these tables is always in a certain format, but needs to be manipulated, by us as operators, into a format which our CAM software can use.

## 12.2 Looking up the catalogue data for the tool

It is quite possible a certain end mill is only designed to machine one specific type of material, or it might be an 'all-rounder'. For each material type with which it can cope, the manufacturer must publish two very important facts about that tool. These are:

| | | |
|---|---|---|
| **Vc** | - | **Cutting speed (metres per minute)** |
| **Fz** | - | **Feed rate per tooth (millimetres per tooth)** |

I have made them larger font and bold because they are important. When you choose a cutting tool, you will need to get in the habit of looking up the data from the catalogue and noting down these values. Probably if you have been using a manual mill, this is not something you paid much attention to before.

One thing to note is these are usually maximum values. The manufacturer is thinking more about industrial efficiency, maximising production speed versus tool wear, minimising cost. As hobbyists or start-ups, we care less about these aspects, but it would be useful to know at least if we are in the right ballpark. We don't need to aim for the maximum but more importantly we don't want to inadvertently operate beyond the recommended limits and break our tools. I can only repeat the message: we don't have a feel for the numbers so it's best we look them up.

These two pieces of 'look-up' data are not very useful to the milling machine controller. We need to use a little mathematics to convert them to data the machine understands and, luckily, someone has been there before in the twentieth century.

There are some formulae which have become 'standard issue' and I need to walk us through these. We will look up the values in the catalogue, enter them into the formulae, and derive some new parameters which we can load into our CAM software which, in turn, will post the G-code to run our mill. It is as simple as that.

This is not a maths book. Therefore, I am not inclined to derive the formulae from first principles, nor prove they are correct. They are not particularly difficult to derive and there are plenty of online resources if you feel the need. But we do need to know and use them. I am going to take it step by step, one parameter at a time, in a sensible order. Luckily the easiest comes first.

## 12.3 Spindle speed in rpm

How fast we spin the cutting tool reflects directly on how fast its periphery moves, and therefore cuts, obviously. Here comes our first formula.

$$n = \frac{Vc \times 1000}{\pi \times D}$$

The value 'D' is the diameter of the cutting tool, and it is expressed in mm. You'll notice the value from the catalogue for Vc is expressed in metres per minute, and this is the standard. The trouble is we are mixing up different units (metres and millimetres), a bit of a no-no in formulae, so a 'fiddle factor' of 1000 is inserted to get round this. But we don't need to worry about any of that if we just use the formula as given, and everything will work out fine. The value 'n', by the way, is the spindle speed in rpm. This is something our CAM software needs to know.

So here is our first problem. Many catalogues just tell you what rpm to use; they have done the calculation for us and put it all into a tidy spreadsheet. That's all well and good, but they are aiming this at their prized industrial customers who run machining centres with spindle speeds way above 30,000rpm. I have seen data for tools cutting aluminium quoting spindle speeds more than 100,000rpm. These expensive machines may have special compressed-air spindle bearings and are 'state of the art' but, back here on Earth, we have a Tormach® 770MX to program and it has two spindle speed pulleys to choose from, so the maximum speed is either 3250rpm or 10,000rpm.

It's more than likely, for cutting Aluminium alloy, the 'book figure' published for Vc will come out too high for our Tormach®, and simply be unachievable.

Here's a real example: Let us look up the figures for a genuine tool, one I have purchased, called the E5H24 sold by Cutwel Ltd (it's a carbide bull-nosed end mill). We will look at the data for side cutting, they call it profiling, there is a handy diagram to clarify what they mean. All the different diameters they sell of this tool have the same value of maximum Vc = 610 m/min. I bought a 6mm one of those, entered the numbers into the formula, and I get a value for best spindle rpm of over 32,000rpm. Whoops. Our machine won't do that. Now what do I do?

As a basic law of physics, the gearing ratio from the motor to the driven spindle affects the torque, in our case the cutting force. For the moment, let us assume we set our machine up with the pulleys set to 3250rpm maximum speed. Although this is the lower speed it will give us better torque. This is how I began running my Tormach®, so it's a good place to start.

So now, forget about the suggested Vc value from a book, let's just use the formula in reverse and effectively 'ask it' what is the best value for Vc I can get out of my Tormach® mill running at 3250rpm? And the answer is the 6mm end mill I purchased will give a Vc value of approximately 61 m/min. That's nothing like the book figure, demonstrating we are operating at machining rates much lower than industrial setups would be able to achieve. But that's fine. It's a thirty-thousand-dollar machine, not a three-hundred-thousand-dollar machine. Dollar for dollar, I guess it's about right.

For small diameter tools, we are limited on what Vc we can achieve, so in this case we will set the speed to our current maximum, which is 3250rpm, as there is certainly no point in going slower. But there is more to consider. The feed rate per tooth was the other very important data in the look-up table. We now need to convert that to a machining feed rate.

## 12.4  Feed rate

Our CAM program is asking us what feed rate in mm/min we want to use for every cutting operation we ask it to calculate. It is referred to as Vf. The feed rate is the rate at which the table moves the workpiece towards the cutter. It is not the same as the 'feed per tooth' quoted by the tool manufacturer. This is where we need the next 'standard' formula, which will derive a value for Vf using the book figure for Fz. Again, if you need to explore the maths behind it, there is an abundance of information online.

$$Vf = Fz \times Z \times n$$

I want to introduce you to another variable in the formulae, called Z. This is simply how many flutes the tool has, so in other words think of it as how many teeth go past the cutting point each revolution. It's obvious why that would affect how much each tooth gets to 'bite off'. The more flutes, the less material each tooth takes a bite of, so that's why Z is needed in the formula.

The reason the Fz parameter is important to the tool manufacturer is simple. If each tooth of the cutting tool is asked to 'bite off more than it can chew', the tool will break. Therefore, it is unsurprising the faster the tool is spinning, the smaller each bite will be. Which in turn means we can feed the material to the cutter faster.

The reverse situation, using our Tormach® at only 3250rpm, we will need to reduce our feed rate so as not to 'break' the Fz 'book figure' limit, causing the tool to fail.

So, example time again, the same tool as before from Cutwel, the 6mm E5H24 end mill. And let us consider we are going to use a maximum speed of 3250rpm (don't forget the book figure wanted over 32,000rpm so we need to figure out the feed rate based on our much-reduced rpm). You will not be surprised to hear the published values of Fz vary with tool diameter. Put simply, the fatter tools are stronger and can withstand taking a greater 'bite'. So, from the data, it says Fz for the 6mm tool is 0.076 mm/tooth. And this happens to be a 3-flute tool. Pop the numbers into the second formula and the answer is Vf = 741 mm/min.

Well, that sounds reasonable enough. Use my little trick; knock a nought off, then divide by 6 and you get approximately 12 mm/sec. I can visualise that and it sounds pretty good. Don't forget this is a maximum figure; it places stresses on each tooth that the manufacturer states the tool will cope with, but no more is permitted. In truth you would probably reduce it a little and see how things go. Walking before we run is advisable. I would enter 420 mm/min into Fusion 360 for the Feed rate for this tool, just for round numbers, and do a test run. I wouldn't use the book figure as my starting point.

As a matter of interest, had we been able to spin this tool at 32,000rpm how fast could we have made the feed rate then? The answer is more than 7000 mm/min. That's 116mm per second, or about 4.5 inches per second. You can begin to see why spindle speed is important. The industrial machinery can 'whizz' through Aluminium like butter when their spindles are running at such fantastic speeds.

But I know you are thinking to yourself, "tell me again why we can't just feed through the STOCK at 4.5 inches per second on our Tormach®". And the answer is, if we did, since our spindle speed is ten times slower, it means each tooth would need to take ten times the amount of STOCK for every bite and that would break the tool. You can see it is totally reasonable running this tool much slower than the 'book' figure, but we must use the formula to ensure we don't exceed the limitations of the tool. In this example, we were physically unable to exceed the 'cutting speed', but we could easily have exceeded the 'feed per tooth' limit. We must be certain not to exceed either.

The question arises, why don't we change the pulley over on the Tormach® and at least run it at 10,000rpm? Before we consider that, there are a couple more things to discuss on feed rate. It is not just a matter of how fast we 'trawl' through the stock, it is also important we take account of how high and wide the cut we are taking is. We will come back to the higher spindle speed later.

## 12.5  Depth of cut, Width of cut

The feed rate is not the only concern here. The depth of cut and the width of cut is also going to affect the strain on the spindle, and on the tool. Remember this Tormach® 770MX only has 1.5 horsepower driving the spindle, so if we demand too much from it, it could 'bog' down and stall completely. So long as we choose sensible depths and widths of cut (don't bite off more than we can chew), then the calculated feed rate ought to be fine (especially if we allow a safety factor of about half as we said), but only so long as we have sufficient power. Some manufacturers' tables will offer depth and width data, and specify limits on those parameters in their catalogues, which are standardised as Ae and Ap. These are limitations on the stresses to which their tools may be submitted but, of course, offer no information about whether your machine could provide the power needed to accomplish such cuts. We will come back to power requirements later.

    Ae  -    width of cut (mm)
    Ap  -    depth of cut (mm)

Looking once more at my example tool from Cutwel, for profile cutting Aluminium, the book indicates that values of Ae and Ap should be expressed as functions of the diameter of the tool. That's hardly surprising, the thicker the tool, the more rigidity.

For taking the side off some STOCK (called 2D profiling), it says maximum Ap = 1.5 x D, and maximum Ae = 0.5 x D.

Using my 6mm end mill, therefore, I could take depths of cut of 9mm and widths of cut of no more than 3mm. The Ae and Ap figures quoted by the manufacturer are limits. But say I needed to cut to a depth of 10mm with this tool, I could do it in two passes, 5mm deep each pass, or maybe if I really wanted to do it in a single pass, say for finishing purposes, it would still be successful if I used a finishing radial stepover with a very fine cut. Like in all things, there is a balance to be had. The manufacturer is suggesting you could use the Ae and Ap limits simultaneously, of course.

The next question to consider, though, is do we have sufficient power to make the cut?

## 12.6 Power considerations and Material Removal Rate (MRR)

I am not suggesting for one moment you will be calculating the power required for each cutting operation. That would be annoying in the extreme, and probably daft. But I would like to look at this aspect, in this chapter; otherwise, we might never truly know how close to the limiting envelope of the machine we will be operating. I think you will find it a worthwhile and interesting exercise. But also, our calculations may indicate we need to back off from the book figures of Ae and Ap simply to reduce the load on the machine, so it doesn't 'bog' down or overheat. Let's try to find out.

First off, how much power do we have at our disposal? The Operator's manual quotes the Tormach® 770MX as having a 1.5 horsepower spindle motor. 1HP is about 745W of power, so we are looking at 1100W. There will be some mechanical and electrical inefficiencies so the cutting power available at the spindle will not be the same as the electrical power the motor is receiving, but that is a factor called 'efficiency' which we can consider separately. Typically, the power at the spindle might be 80% to 90% of the electrical power used by the motor.

OK, so second off, how much power do you need to cut things up? It is surely an interesting question in its own right - and warrants some delving. A colleague once proffered a rough 'ball-park' figure when he told me,

"It takes one horsepower to remove one cubic inch of low carbon steel per minute". That's fascinating but, upon hearing it, I immediately wanted and needed to know more. I now wish to examine how we can calculate the finer details more scientifically. I am anticipating this will help us understand the subject better and will hopefully reveal what sort of material removal rate we can expect from the Tormach® 770MX. That would be useful, I trust you agree.

It transpires some clever gentlemen in the early days of manufacturing did some research into how much energy is required to cut various materials. Certainly, the tensile strength of the material is a factor in how hard it is to cut, but that parameter alone doesn't reflect the full story as there are added effects from the physics of the cutting tool rake, how much energy goes into chip generation, tool sharpness, and what sort of cutting process is being used. This is all beyond our needs here but suffice to say we can look online these days and 'grab' a handful of the data they found out.

One parameter they came up with is called the Specific Cutting Energy (SCE) and you can think of it as the amount of energy it takes to remove 1 cubic millimetre of material, in joules.

Here are some numbers I found on the web. They are approximate as every alloy is different, but it is a useful guide. They are also very subjective. Even the sharpness of the cutting tools will affect the cutting energy required, as will the rake of the cutting edge. Also, the amount of energy varies slightly depending on the 'bite size' of the machining, (bigger bites will tend to save energy overall). There is no 'right' answer. That is probably why, when you purchase metal stock, the supplier doesn't quote you any SCE value with the sale. All they might suggest is 'good machinability' just as they might say 'good weldability'. Pretty useless in a scientific way, but very useful in a practical way.

- Mild steels         SCE = 2.0 to 2.8   (joules/mm^3)
- Tool steels         SCE = 2.5 to 4
- Stainless steels    SCE = 2.8 to 3
- Aluminium alloys    SCE = 0.6 to 0.8
- Brass               SCE = 1.5 to 2
- Cast Iron           SCE = 1.0 to 1.5

You can glean, from the data, it takes about three times more energy to machine a lump of mild steel, than it does the same size lump of aluminium alloy. Put another way, it would take three times the power if you sought to do it in the same amount of time.

I want to test the 'one horsepower / one cubic inch' statement, for mild steel, to see if the figures I have found on the internet are reasonable. One cubic inch is 16,387 cubic millimetres (simply 25.4 cubed). Let's use the newfound data and select a value of 2.8 joules per cubic millimetre. That would mean a cubic inch would require 45,884 joules to be fully machined into swarf – an odd thought to put it mildly. My colleague suggested if we wished to complete this in one minute, we'd need one horsepower. Using 45,884 joules in a minute equals 765 joules of energy every second, which is the same as saying 765 watts, since the rate at which you expend energy is the power at which you are working. Usefully, this little excursion has confirmed a link between what my colleague told me and the data I found independently on the internet. Recall a horsepower is 745 watts so I am going to call that a 'bullseye'. The SCE data appears to be valid.

As an added complication, the SCE is also known as the 'Specific Cutting Force' and given the label 'Kc'. And the units are 'newtons per square millimetre', and sometimes expressed as megapascals (MPa) which are units of Pressure. This seems odd at first sight, but the vagaries and intercorrelation of the scientific units does indeed mean that 'joules per cubic millimetre' and 'pressure' are indeed equivalent things, as strange as that sounds, albeit there is a 'fiddle factor' of 1000 due to the mixing up of millimetres and metres (Divide 'Kc' values, which are in MPa, by 1000 to obtain the SCE number in joules per cubic millimetre). If you can get your head around the initial strangeness of the concept, it is the hardness and tensile strength of the material which is the main determinator of how difficult it is to machine, which is also a measure of how much pressure it can resist.

Now we have some useful ballpark data for the energy required to remove a cubic millimetre of material, we need to figure out a way of calculating the material removal rate (MRR) the milling machine is achieving. That is straightforward. Think of an 'area' of material being blasted out of the way as the cutter comes through. The 'area' is simply Ae x Ap (width of cut multiplied by depth of cut). The 'volume' of material we sweep up in the process is that 'area' multiplied by the distance we feed the cutter through in a second (Vf, the feed rate). As we always measure feed rates in millimetres per minute, we need to divide it by 60 to come up with an MRR per second.

$$MRR = \frac{Ae \times Ap \times Vf}{60}$$

And the rate of removal multiplied by the amount of energy needed to remove one unit of volume (SCE) therefore gives us the power required to do it. This is the 'cutting power' or Pc.

$$Pc = \frac{Ae \times Ap \times Vf \times SCE}{60}$$

That is the power required at the cutting face. To calculate the power required by the motor, we would need to factor in inefficiency. Typically, there might be 20% losses. That would mean increasing the answer we get for Pc by 25% but let's not concern ourselves too much about 'motor efficiency' and 'cutting efficiency' now. Accepting the 'mechanical power' we can extract from the spindle is less than the 'electrical power' the motor soaks up, I want to use this new formula to work through some real-world examples to get just a rough idea of how much power we are going to ask the Tormach® to deliver. Just using blunt tools would be enough to increase the actual value of SCE for a material, so it's all a bit subjective anyway.

## 12.7 Power calculation examples for the Tormach® 770MX

Let's continue this investigation using a couple of typical examples with real tools, using their actual data from the manufacturer's tables. And we will use the formulae above to ensure we don't bust any limits of the tools.

In the examples we will cut some aluminium alloy with a tool designed for that, a 6mm 3-flute carbide corner-radiused (bull-nosed) end mill. Then we will cut some mild steel using a different tool, a 6mm 4-flute carbide end mill specifically designed for machining steel. In each case, the tool data derived from their respective manufacturer and will inevitably be typical of those sorts of tools.

Example 1. ALUMINIUM

- Aluminium alloy (assume SCE is 0.8 joules/mm^3)
- Tool data D = 6 mm
- Tool data Z = 3
- Tool data Vc (max) = 610 m/min
- Tool data Fz (max) = 0.075 mm/tooth
- Tool data Ap (max) = 9 mm
- Tool data Ae (max) = 3 mm

What we want to do:

- Ap = 9mm
- Ae = 1mm

Our calculations:

- n = 3250 rpm
- Vf = 731 mm/min
- MRR = 110 mm^3/sec
- Pc = 88 W

Example 2. MILD STEEL

- Mild steel (assume SCE is 2.8 joules/mm^3)
- Tool data D = 6 mm
- Tool data Z = 4
- Tool data Vc (max) = 95 m/min
- Tool data Fz (max) = 0.030 mm/tooth
- Tool data Ap (max) = 6 mm
- Tool data Ae (max) = 0.6 mm

What we want to do:

- Ap = 6 mm
- Ae = 0.6 mm

Our calculations:

- n = 3250 rpm
- Vf = 390 mm/min
- MRR = 23 mm^3/sec
- Pc = 65 W

Both these examples appear not to be taxing the Tormach® too much in terms of the power needed to complete the operation. The mild steel example is taking two-thirds the power, but the material removal rate is only about a fifth, so you would need to spend five times as long doing the job in mild steel. In doing so, you'd use about three times the energy and that's because it's about three times as hard to machine. Now you can see the importance of the SCE values.

Let's do a third example. This time we will ramp up the pressure on the Tormach®, choose a larger diameter cutter with more flutes, and cut some stainless steel.

Example 3. STAINLESS STEEL

- Stainless steel (assume SCE is 3 joules/mm^3)
- Tool data D = 12 mm
- Tool data Z = 4

- Tool data Vc (max) = 95 m/min
- Tool data Fz (max) = 0.045 mm/tooth
- Tool data Ap (max) = 12 mm
- Tool data Ae (max) = 1.2 mm

What we want to do:

- Ap = 12 mm
- Ae = 1.2 mm

Our calculations:

- n = 2520 rpm (since we would exceed Vc if we drove it faster)
- Vf = 454 mm/min
- MRR = 109 mm^3/sec
- Pc = 327 W

Well, so far, so good. It is looking like we should be able to do some useful work. Even if we factor in some inefficiencies, the electrical demand on the main spindle motor looks to be about half of what's available. There are so many other examples to play around with. Try using a facing tool with much greater diameter. It might have many indexable inserts (equivalent to flutes, Z in the formula). That will have an effect. Or you could take out all the inserts except one, thereby creating a 'fly-cutter' and see what effect that has on the power requirements. Now you are equipped with the formulae, I will leave you to experiment with the possibilities in your own time. Let's move forward and talk about torque limitations.

## 12.8 Torque considerations

We said at the outset, it was very difficult for a CNC mill operator to have mechanical 'feel' for how hard the machine is working, compared to someone manually winding the handle on something like an old-fashioned Bridgeport. It stands to reason the CNC operator is 'distanced' when he is sitting at his computer working through the CAM software.

We also briefly spoke about not wanting to 'push' the mill so hard the spindle motor gets 'bogged' down and stops. When you think about it, that would happen if the mechanical force required to cut the material is greater than the turning force of the spindle motor. In short, we need to consider what torque we have available and what amount we think we will need.

## 12.9 Torque available

How much torque is available at the spindle? The maximum power a motor can safely produce, occurs when the maximum rated torque turns the motor at rated speed, while the windings are taking maximum rated electrical power. And there is a formula for that. It has to do with the physics of the relationship between power, torque, and angular velocity, with the latter expressed in 'scientific' units called radians. This formula will prove to be useful.

Now I am absolutely no expert on motors, and their science. The electrical engineering behind three-phase induction motors is complicated enough when they are running on their normal 50Hz or 60Hz mains and gets a whole order of magnitude more complex when you feed them a variable frequency to control their speed using a Variable Frequency Drive (VFD). So, I am going to have to 'skirt' around the fundamentals just to see if we can get some appreciation of the numbers.

When you control motor speed with a VFD, you can operate it beyond the normal conditions printed on its nameplate. Specifically, when the frequency increases above the normal mains frequency, there is a fall-off in output torque from rated torque.

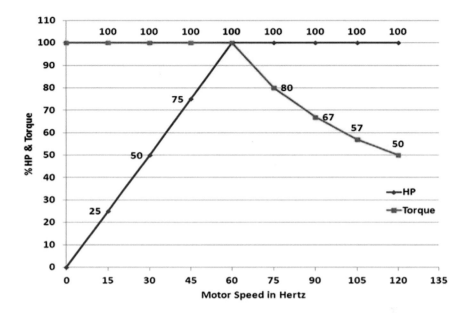

**Relationship of horsepower, torque, and speed under VFD control**
Copyright© graphical information courtesy of Joe Evans

The rated torque of the motor is only available up to its 'nameplate' operating frequency and then decreases beyond that. From physics lessons in school, Torque and Power are related by a simple equation:

Power = Torque x Angular velocity

$$P = T \; x \; w$$

where angular velocity is basically the speed of rotation, but measured in radians per second, rather than just simply rpm. (Radians are simply a scientific way of measuring angles. In a whole 360-degree rotation, there are just 2π radians, so a radian is about 57 degrees, but that truly doesn't matter here.)

$$w = \frac{2\pi \; x \; n}{60}$$

If 'w' is the angular velocity and 'n' is the rpm, it is just a conversion of units from revolutions to radians. Anyway, combining the two equations above into one, we can figure out what the maximum torque available is:

$$T = \frac{Pc \times 60}{2\pi \times n}$$

'Pc' denotes 'cutting power.' The 60 Hz section of the nameplate on the motor of my Tormach® 770MX says it is rated at 900 watts when running at 3360 rpm. (It is a 2-pole motor so it would run at 3600 rpm if it wasn't loaded but slows down a bit as it gets loaded up, reducing to 3360 rpm at its rated load and that's when it is demanding 900 watts. I believe this is called the rated 'slip' speed.

T = (900 x 60) / (2π x 3360)

I come out with an answer of 2.6Nm.

Keep in mind Tormach® publish the spindle power to be more like 1,100 watts and I understand this has to do with some 'clever' over-driving of the motor by the VFD, but I am going to remain conservative and not consider this aspect, since I don't have any data on the efficiency of the entire system either. There will of course be losses for which we have made no allowance.

So, as far as I understand these things, the motor will produce a maximum rated torque of 2.6Nm when it is run from zero speed up to 60Hz supply frequency. Typically, the VFD will be able to take the supply frequency up to twice rated, so 120Hz in this case. And in doing so, the power will not increase beyond rated power due to the physics of the ever-increasing 'back emf' voltage subsequently reducing the current, and therefore also the torque. That's what the graph is showing.

The spindle on the Tormach® 770MX is not a direct-drive design, instead it is belt-driven via a pulley system offering two manually selected options. The 'low speed' range is a 1:0.45 ratio and designed to give a maximum spindle speed of 3,250rpm, whereas the 'high speed' range is a 1:1.5 ratio and designed to give a maximum spindle speed of

10,000rpm. (These are approximate ratios derived simply from measuring the diameters of the four pulleys with a steel rule).

Interestingly, when you use these ratios to calculate what the motor speed is doing at the respective maximum spindle speeds, it works out at about 7,200rpm on the 'low speed' ratio and 6,670rpm on the 'high speed' setting. Recall this is a 60Hz rated motor and I mentioned we can expect the VFD to run the motor up to twice its rated speed, by increasing the supply frequency to 120Hz, so these maximum speeds, then, probably relate to that scenario. So why are they different? I am guessing, but I don't really know (you must give me full marks for honesty here), the 'high speed' setting offers less torque at the spindle, so will tend to slip more than the 'low speed' setting. At 60Hz, in 'high speed' setting, the motor speed would be 3335rpm – half of 6,670rpm - (which is remarkably like the rating plate figure of 3360rpm), but in 'low speed' setting we would infer the motor speed to be 3600rpm (which is zero slip, so unlikely in practice).

Apart from some losses due to inefficiency, power is conserved across pulleys with a gear ratio; it is the torque which changes in an inverse relationship with the speed. In simple terms, if the pulley system increases the speed, there will be a corresponding reduction in torque. I previously estimated the torque available at the motor to be 2.6Nm. Therefore, it is reasonable to imagine the torque available at the spindle to be 5.8Nm in 'low speed' mode, at up to half the maximum speed (so up to 1,625rpm). I divided the motor torque by the gear ratio to get this value. Anything above 1,625rpm, and the torque would fall off, just like the curve on the graph, to approximately half its best value by 3,250rpm. (We are assuming the best torque is available from the motor when it is run at 60Hz).

Conversely, in 'high speed' mode, the torque at the spindle would be 1.7Nm up to 5,000rpm (or whatever spindle speed we get from driving the motor at 60Hz allowing for some slip), and then tail off to about half that value by 10,000rpm. So, you can understand there is an appreciable reduction in available torque in 'high speed'.

I approached Tormach® for some data on their 'torque curves' to see if my amateur calculations were anywhere near the ballpark. I am rather delighted to say they were. Of course, in the real world, the power and torque available at the spindle are never going to exactly imitate the graphical presentation shown above; the theory behind such a curve comes with many simplifications. Andrew Grevstad, Business Development Director at Tormach® in the USA, has kindly permitted me to reproduce their test data for operating a 770M machine (same spindle motor as the 770MX) in the 'high speed' mode.

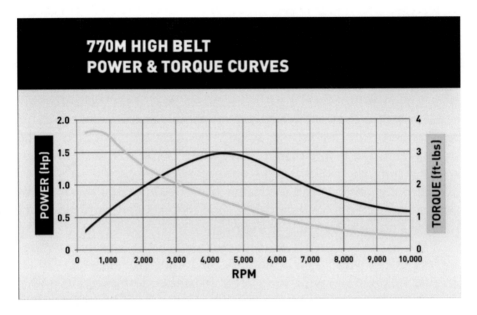

**Tormach® power and torque curves**
Copyright© chart courtesy of Tormach® Inc.

Obviously, for those of us preferring to use metric units, this graphical presentation of data delivers us the added complication of unravelling the North American ones on each side. A newton-metre of torque (Nm) is about 0.75 of a foot-pound (lb.ft.) and 745 watts of power (W) is about one horsepower (hp). I shall leave you to 'pencil in' your own units.

Interestingly, I reckon the green line cuts the 5,000rpm point at about 1.75Nm of torque, thus reinforcing my earlier calculations. I don't have the equivalent graphical data for the 'low belt' configuration. As far as I can surmise, though, the power and torque curves for this should be similarly shaped; it is the same motor, with just a gear change. But in the 'low belt' regime, I would expect there to be slightly less induction motor 'slip' so the peak power point might be expected nearer to the centre of the curve, the 60Hz point if you will. So let us summarise what we think we have at the 'peak power' point for each belt setting. Note the 'real' data suggests peak power happening at about 4,500rpm in 'high belt' configuration. I am suggesting it will occur somewhere near 1,625rpm in 'low belt' setting (half of 3,250rpm). Using the pulley ratios determined earlier, I have drawn up a chart of the approximate torque expected to be available at the spindle, for the peak power speed, and the maximum speed, for each of the two belt configurations (hi/low). It is only an estimation, but it gives you a rough idea of where things stand.

|  | Gear ratio | Peak Power | Torque | Max speed | Torque |
|---|---|---|---|---|---|
| Low Belt | 1:0.45 | 1,625 rpm | 6.6 Nm | 3,250 rpm | 2.0 Nm |
|  | Derived motor torque >>> |  | 3.0 Nm |  | 0.9 Nm |
| High Belt | 1:1.5 | 4,500 rpm | 2.0 Nm | 10,000 rpm | 0.6 Nm |

Looking at the graphical data from Tormach®, it looks like the peak power (which is 1.5 horsepower or 1.12kW as we already established) falls off at maximum speed to about 450 watts (or 0.6 horsepower). Therefore, there may well be some advantage to operating the machine at approximately half the maximum speed, of whichever belt setting is being used, to get best power and torque combination.

Conversely, if the material being machined is not particularly difficult to cut (low SCE number), then greater speeds (rpm) will obviously offer the opportunity to increase the feed rates, subject to staying within the 'feed rate per tooth' data (Fz) for the cutting tool in use. In other words, if the material is easy to machine, then you can plough ahead and get the job done quickly, whereas if it is difficult material to deal with, then you might need to reduce the speed to half the maximum, in order to improve the power and torque available and this will most likely necessitate reducing the feed rate too, to accommodate the tool limitations.

And clearly, the 'low belt' ratio gives the most torque available of all, should you need it. So how much torque do we actually need?

## 12.10 Torque required

That last equation, which demonstrates the relationship between power and torque, is very useful. Because for every power calculation we make - and we worked through three examples earlier - we know what the rpm is going to be, since we chose it.

In all the calculations so far, we have allowed for 'how hard the material is to machine' by using a value for SCE, we have gone through all the tool limitations from the manufacturer to come up with a sensible cutting regime, and we ended up with a required power figure. Since we know what speed that power is being consumed at, we can use the formula above to directly work out the mean torque required to do the job.

Here are the figures summarised from the earlier examples:

| | Material | Cutter | Speed | Feed rate mm/min | MRR mm^3/sec | Power W | Torque Nm |
|---|---|---|---|---|---|---|---|
| Example 1 | Aluminum alloy | 6mm, 3 flute | 3250 | 731 | 110 | 88 | 0.26 |
| Example 2 | Mild steel | 6mm, 4 flute | 3250 | 390 | 23 | 65 | 0.19 |
| Example 3 | Stainless steel | 12mm, 4 flute | 2520 | 454 | 109 | 327 | 1.29 |

All three examples above, assumed we had the machine in the 'low speed' mode, so the maximum speed available was 3,250rpm. Even allowing for the 'tail-off' on the graph, it is clear all three examples present no difficulty for the mill, in terms of mean torque required. When we first discussed machining the aluminium alloy, it was noted the manufacturer's tool data implied much higher speeds would be beneficial, so let us now consider switching the machine over to the 'high' speed setting and noting the effect.

## 12.11 What about the 10,000rpm option on the Tormach® 770MX?

For completeness, I wanted to look at what effect the option of switching to the 10,000rpm setting is on the Tormach®. It is not a thing you can do at the flick of a switch unfortunately. You must loosen the motor mount and change the drive belt over to the alternative pulley ratio. You must also remember to enter on the PathPilot® touchscreen that you've done that. Otherwise, it wouldn't know.

In fact, it is only the first two of my three examples which would benefit from doing this, as the tool data suggests we cannot machine the stainless steel with the 12mm tool at such a fast rpm. So let us go through the formulae for the first two examples again, this time accepting our maximum speed is now 10,000rpm.

## Example 1. ALUMINIUM (10,000rpm available)

- Aluminium alloy (assume SCE is 0.8 joules/mm^3)
- Tool data D = 6 mm
- Tool data Z = 3
- Tool data Vc (max) = 610 m/min
- Tool data Fz (max) = 0.075 mm/tooth
- Tool data Ap (max) = 9 mm
- Tool data Ae (max) = 3 mm

What we want to do:

- Ap = 9mm
- Ae = 1mm

Our calculations:

- n = 10000 rpm (the maximum possible)
- Vf = 2250 mm/min
- MRR = 337 mm^3/sec
- Pc = 270 W
- T = 0.26 Nm

## Example 2. MILD STEEL (10,000rpm available)

- Mild steel (assume SCE is 2.8 joules/mm^3)
- Tool data D = 6 mm
- Tool data Z = 4
- Tool data Vc (max) = 95 m/min
- Tool data Fz (max) = 0.030 mm/tooth

- Tool data Ap (max) = 6 mm
- Tool data Ae (max) = 0.6 mm

What we want to do:

- Ap = 6 mm
- Ae = 0.6 mm

Our calculations:

- n = 5040 rpm
- Vf = 605 mm/min
- MRR = 36 mm^3/sec
- Pc = 101 W
- T = 0.19

For both examples I have tacked on the torque figure at the bottom and, interestingly, these have not changed at all from the earlier calculations despite the fact we are running at a much faster speed. In each case, what has changed is the MRR and Power consumption have increased so we are getting the job done faster. But the actual job wasn't harder to do.

Bearing this in mind, the third example I used was cutting stainless steel and we could not do the job any faster as we were limited by the tool data. Remember we had to reduce the spindle speed to something like 2,500rpm. But the mean torque was already up at 1.29Nm for this job. As we have been shown from the data afforded us by Tormach®, below about 5,000rpm, we have comfortably at least 2.0Nm of torque available so it appears this job can be accomplished in either of the speed settings. And a quick check of the power curve also reveals either configuration will be acceptable.

Running in high-speed mode will offer less, nonetheless sufficient, torque. On the other hand, the motor speed will be lower because of the pulley ratio, so you could expect less noise (and presumably less motor bearing wear).

What is clear, though, for the first two examples, we did get both jobs done faster by using the 'high speed' setting – you can see this is the case by referring to the MRR figures. But, for tinkering with hobbyist projects, or doing some one-off prototyping, where speed is not our main concern, then the higher torque margins might be preferable using the 'low speed' setting. On the other hand, in many cases you will not need the higher torque afforded you by the low gear, so you may prefer to 'run quiet'. In the case of volume production, when the tool/material combination permits, then getting the job done faster will surely improve the bottom line.

Even making allowance for inefficiencies, it is clear these examples fall well within the design envelope of the Tormach® 770MX. Equally clear, it would be no great hurdle to create some badly designed CAM operations using tools with perhaps larger diameters, different number of flutes, too greater depths or widths of cut, to push the mill too hard. I hope this chapter demonstrates how important it is to be aware of the limitations of the tools and the machine, and the only real way of knowing how close you are operating to those limits is by considering the mathematics. The formulae shown are standard in the CNC industry and equip the operator with the means to have much greater 'feel' for the numbers, rather than just guessing.

That said, in the hobbyist world, there will always be times when you are doing a one-off job and you will have become so used to the sorts of feeds and speeds your own tools are capable of, you can and will make sensible guesses for feed rates. That's fine, but the same cannot be said for drilling holes.

## 12.12 Drilling holes

So, for milling cutters travelling sideways, we can generally 'get away' with things by 'erring' on the less ambitious side of the cutting regime. Indeed, we may well know what the tool limitation is by looking up the data in the catalogue, but in the hobbyist world we don't have any great need to run the tools at their maximum stress levels. It is easy to 'guess' low. Where I have found this is not the case is with drilling holes. Twist drills are not very forgiving if you try to push them too hard; they are not that strong 'end-on'. Particularly very small diameter ones. If you think about it, whether we are drilling a hole with a hand-held power drill, or doing some more accurate drilling using a drill press, or even the quill feed on a manual mill, we are constantly 'feeling' for how well the drill is cutting. It is a natural human response to apply just the right pressure to keep the drill focussed on cutting the hole without forcing it and snapping it off, and without 'letting off' too much causing the tool to rub, squeal, and get hot. But if you drilled a hole in some steel using a drill press, and then I asked you,

"What feed rate do you think you just used?", you probably couldn't think of a number even close. You just 'eased' that drill down so it was always cutting 'just nicely', using your inherently sympathetic skill.

Unfortunately, that 'built-in' skill is of no use to us if we must figure out what 'plunge rate' value to enter into our Fusion 360 CAM software to program this drilling operation. We need a number. We can't program the computer with 'just do it carefully please'.

In my first attempts with the Tormach® 770MX cutting some 2mm diameter holes in aluminium alloy, I didn't realise just how sensitive the situation was, and I broke some drills. Some rather expensive ones it must be said, which made me sit up and investigate this much more carefully. It was the equivalent of me at the drill press, simply pulling far too hard on the lever and expecting the drill to cut faster; it simply cannot and will instead snap off leaving the broken part of it lodged inside the work, which then becomes scrap. I genuinely thought I was being cautious but, in hindsight, my guessing the parameters was insufficiently accurate. We absolutely must do the calculations using the manufacturer's tool data to ensure we do not ask too much of a drill.

What are the variables? Well, the diameter of the drill is clearly a known value. The manufacturer will once again publish a cutting speed, Vc, and this refers to the fastest point of the drill, so therefore at its periphery. There is one further data parameter we need about the drill and that is something called the 'feed per revolution'. This is found, again, in tabulated form in manufacturers' catalogues, usually online. And it is specified according to which material we wish to drill into, so look out for another PMK chart, with the now familiar colour coding. But note, 'feed per revolution' is a new type of parameter, sometimes named 'feed per rotation', different to the 'feed per tooth' value given in the milling cutter charts. Again, for completeness, here is the notation formally:

$F_n$   -   feed per rotation from manufacturer (mm/rev)

So, the whole calculation process goes something like this: Starting with the parameter Vc, we can use the formula already established to calculate a maximum spindle speed. This part is no different to the calculation for any milling cutter. Next, we would need to calculate how fast to drive the drill downwards into the material. And for this, I need to introduce you to a new formula to calculate the vertical feed rate (downwards to drill the hole) which is as follows:

$$Vf = Fn \; x \; n$$

In other words, for drilling, we look up the manufacturer's values for Vc and Fn. From Vc, we can use the first formula to find a suitable value of n, the rpm. That value may be limited by the tool, or it may be limited by our machine's maximum. The CAM software needs to know the rpm. Once we have the rpm value, we then use the last formula to calculate the feed rate. That is the figure which the CAM software also needs to know, so it can drive your drill into the material without breaking it.

Now with drill data, there are lots of extra notes and warnings concerned with how deep you are planning to drill. The problem with deep drilling is the swarf clearance (chip evacuation) becomes more difficult the deeper into the hole the cutting edges are working. It is widely known to be a particular problem on lathes since the drill isn't rotating - the workpiece is instead - so chip evacuation is less assisted. I would suggest the value for drill feed rate derived from the formula is a maximum and you should work back from that figure if there are any complications. Any hole which is deeper than approximately three times the diameter is considered a complication. From drill manufacturers' datasheets, ensure you have read all the notes at the bottom.

Another factor to be aware of is the different ways to drill a hole. There are special macros for different drilling techniques which allow you to peck at the hole in stages, fully retracting between each peck, or partially retracting between each peck, or not pecking at all and drilling in one go. This is all evident when you fill in the data in the drop-down menus within the drilling operation inside your CAM software.

This needs practice, and experimentation, and getting used to the different ways of drilling. In my earlier example, when I was going through a stage in my learning and breaking my 2mm drills in aluminium, as soon as I did the proper calculation using the manufacturer's tool data, and programmed my CAM correctly, I instantly solved the problem of my drills snapping off midway through a cycle. It was a 'lightbulb' moment for me.

The moral of the story is the CNC machine has no 'feel' at all. In the case of drilling feed rates, it's virtually impossible to guess what's right. So, don't try to, especially using smaller sized drills.

When it comes to sourcing twist drills for the CNC machine, any decent-quality drills will suffice. To ensure high accuracy of the hole location, it is important to only use sharp drills, as this negates any tendency for sideways wander, which is often a cause of drill breakages. Most 'expert' resources I have read, suggest always spot-drilling before drilling with a twist drill, although I am aware of some modern drills, which are 'material specific', being advertised as 'self-centring' and thus, in theory, not needing this precaution.

If possible, I always use ER20 collets to hold my drills rather than resort to drill chucks. This is because the drill chuck only has three jaws and these are concentric within 'less than perfect' tolerances, whereas the ER20 collet concentricity is far more precise. Also, the collet has a better all-round grip of the drill, so it can't be pulled from the holder and spoil the tool length data. I have found whichever way you choose to hold the drills, for the smaller sizes, there are some exceptional quality carbide drills available, manufactured on a shank which has a larger diameter than the tool itself. For example, I have some 1mm carbide drills which are formed on a 1/8-inch shank, which is a bit of an odd combination of units I must admit, but they are easy to hold in a collet and work brilliantly. The supplier for these was APT, whom I have listed at the back of the book.

# Chapter 13    Work holding

13.1  Machine vice

I already mentioned I bought the Tormach® '5 inch' vice and it comes with its own mounting bolts and tee nuts. The hex-head bolts are imperial sized so you will need a suitable set of Allen keys. The thread is 1/2-inch UNC x 13 and the Allen key which fits the cap head of these bolts is 3/8-inch.

**Tormach® 5-inch machine vice**
Copyright© photograph courtesy of Tormach® Inc.

Soon after I began using my mill, I realised there would be times when having a second identical vice might come in handy. I ordered another '5 inch' Tormach® vice. There is easily sufficient space on the 770MX table to fit both simultaneously. In fact, in each package, they not only include four mounting bolts and 'L' shaped clamps, but also add a flat metal bar with a single hole. When you only have one vice, you always wonder what this extra piece is for. Cleverly, when you take delivery of your second vice, and consequently now have two of these pieces, it dawns on you what they are for. By using the two metal bars between the vices, you can clamp down in between them with just two bolts, and in the process manage to get the two vices closer together. The pair of vices are therefore held in place with just six bolts, which is plenty. There is still room on the table to secure the electronic 'tool offset' probe as well.

Setting up these vices in side-by-side parallel is simple and using a dial indicator secured to the spindle, or in a toolholder in the spindle, you can run along the faces of the rear (fixed) jaws to set them parallel and exactly in-line.

What you cannot do so easily is ensure the heights of the horizontal ground surfaces of each vice are the same. I used my Passive probe tool to probe the heights of each vice and found one was 0.020mm higher. So, I purchased steel shims from RS Components Ltd. and fitted them beneath the lower vice to cancel out the difference, and it worked well.

Then recently I was looking at a YouTube presentation and came across a suggestion, which had not occurred to me, and which would solve the slight mismatch with no outlay. Assuming you are trying to hold a long piece of stock which stretches across more than the width of the two vices, then all you need to do is set parallels on the mill table itself instead of the base of each vice, thereby eliminating any error from the vices.

13.1.1 Vice soft jaws

Tormach® sell on their main USA based website, steel, and aluminium, soft jaw options for their vices. I haven't had the opportunity to use these yet but look forward to the time a project comes my way which will require their inclusion. I would think it is worth considering getting some of these in stock for when that special job crops up. To cut shapes into your soft jaws to hold flipped parts ready for their second operation needs careful and clever use of your CAM software and there are plenty of examples of people doing this on YouTube. When you are trying to clamp a PART into soft jaws, it is important the moving vice jaw and the fixed jaw do not meet, so that full clamping force is available. So, when designing the cutting of the special shape in the jaws, in the CAM design, you set a notional gap between the jaws. When it comes to the machining, you simply place a parallel, of the same value as the chosen gap, between the jaws and tighten the vice onto it. In this way, the special shape cut into both jaws accurately reflects the shape of the PART, only when that same gap is correct.

## 13.2 Horizontally mounted 3-jaw chuck

There are times when you will need to hold a piece of round bar, or tube, vertically. You could use a vee-block in a vice to hold round bar, but another option is a 3-jaw chuck secured flat to the table. These are readily available and competitively priced, usually with three 'through-holes' for securing to a table with bolts and tee-nuts.

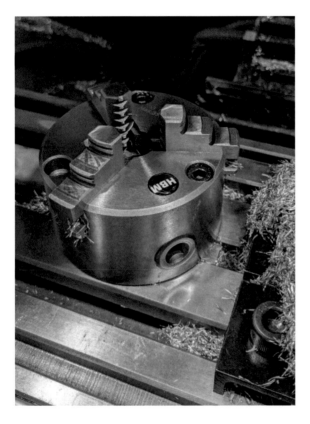

**Three jaw scroll chuck bolted to table**

I purchased one very early on and use it regularly. In fact, the three through-holes do not line up with the table slots on my mill but, of course, I am able to just use two bolts, which I realise isn't perfect, but it does work. Ultimately, I plan to make an intermediate fixing plate to solve the problem. It does not matter about centring of the chuck, since you can probe the position of a round boss automatically using your work offset probe. Position the chuck anywhere on the table, tighten a known straight piece of material into the jaws of the chuck, then probe that to find the vertical centre line of the chuck, noting once more the repeatability of these sorts of scroll chucks is poor compared to a collet system.

## 13.3  microARC4 '4th axis'

The 4th axis unit supplied by Tormach®, whilst it comes under the heading of 'work holding', is a major upgrade to the capabilities of this milling machine and discussed separately in the book. However, it is perhaps worth mentioning here, it has two work holding options, both of which are included in the set, a 3-jaw chuck and a backplate with a built-in ER40 collet chuck. I believe the standard ER collet system goes up to size ER50, but these are seldom seen. ER40 is, conversely, a common size of collet chuck for work holding and the full range of collet sizes, which fit it, goes from 3mm to 26mm. Tormach® do not supply any ER40 collets with the microARC4 unit, so I suggest purchasing a full set.

## 13.4  ER40 collet chuck

For an extremely competitive price, you can purchase a fixed ER40 collet chuck. They come with a square base, or a hexagonal base. I bought a square base version and this clamps into a machine vice easily and squarely by default, which is super useful. As there is already an ER40 collet chuck on the 4th axis unit, buying a full set of collets is made even more relevant and justifies the expense.

**ER40 collet chuck resting on parallels in the machine vice**

The only complication I found is the machine vice has a large side-to-side gap in the centre, and the collet chuck is not able to reach between the horizontals, in fact falling through the gap, so I used two parallels laid flat, to ensure the chuck stands vertically. I used two parallels which didn't completely fill the gap between the vice jaws, to ensure coolant could drain through the centre onto the table and return to the sump.

## 13.5 A simple 'twin operation' work holding example

Continuing with the theme of my ER collet chuck mounted squarely in the machine vice, I would like to describe a particular job I became involved with to demonstrate how much 'thought' input is required each time you must design a 'fixturing' solution. Depending on the part being manufactured, the way in which you sequence the different operations is purely a human decision. Consequently, there are no right answers.

So, I designed a part which was based on a 3/8-inch socket at one end, and had a particular functional shape the other, all made from one piece of metal. I decided to manufacture the 'socket' end first, using the 3-jaw scroll chuck. The vertical centreline of this chuck was measured by probing with the Tormach® Passive Probe, against a known piece of round stock secured in it. The repeatability of this centre is not particularly accurate, by the very nature of scroll chucks, but is good enough for the first operation, since the first end of the PART is smaller than the diameter of the STOCK, so even if the piece of stock is not exactly centred, or even vertical for that matter, the first operation would still produce an accurate part.

The second operation (the other end of the part) should exactly match the first operation in the X and Y axes, so accurate positional fixturing must be achieved. For this, I used the ER40 collet chuck mounted in the vice just like in the photograph. The 3-jaw chuck secured directly to the table and the ER40 collet chuck mounted in the vice were placed side-by-side. In this way, I could achieve a mini production line and manufacture the two halves of the part, one per fixture, so that after running a combined Op1/Op2 program, I would produce a finished part each time, then move the half-finished part from the 3-jaw chuck to the ER40 collet, place some fresh cut stock into the 3-jaw chuck, and repeat.

In my setups for Op1 and Op2, I used G54 and G55 respectively. To probe the centreline of the ER40 collet, I used a piece of 20mm round bar in a 20mm collet. This is very repeatable, much more so than a 3-jaw chuck would ever be, so I could be sure the 'other end' of the part would be manufactured exactly concentrically with the first operation. The exact length of the part (it is a 3/8-inch socket tool) is not crucial so I could accept a less stringent tolerance in the Z direction.

**Probing the X and Y values for G55 using a straight piece of round stock**

Now let's now discuss setting the Z heights in both the G54 and G55 workspaces for this job. The finished PART is 32.0mm long. And I aimed to cut a piece of metal STOCK on the bandsaw which was 33.5mm, allowing for a 'rough' cut and the possibility of the cut not being exactly square to the bar. If I was making a huge number of these, I would try to be more frugal, but this was my first attempt at 'production of a part' so stay with me, please.

The STOCK needed to protrude above the three jaws of the chuck sufficiently for the first operation machining to be able to reach the metal without crashing into the jaws. Because of this, I had to make a small round spacer from steel, for the stock to sit upon. I touched onto the top of the spacer with the Passive Probe to zero my Z value for G54 workspace. In the Op1 setup within my CAM program, I made sure the origin in G54 was the centreline of the STOCK (and the PART), with the Z zero position on the bottom of the STOCK. In this way, when I placed the piece of rough-sawn STOCK into the chuck jaws it rested on the spacer I had made, where I had already told the machine Z was zero. In Fusion 360 CAM, I made sure my 32mm PART resided within the

33.5mm STOCK with 1mm spare at the bottom, and 0.5mm spare at the top. The first task in Op1 would be to face the top surface by taking off 0.5mm, and this assumes the piece of rough-sawn STOCK was cut accurately. I allowed for the case of it being a bit longer than planned by facing off the top starting at 34mm high. If the piece was not so long, then the first pass would be cutting 'thin air'. No big deal. If after facing off to my planned 33mm (that's the 32mm part plus the 1mm 'spare' metal beneath it) the surface was not fully smooth, then I had made an error and sawn the piece of STOCK too short. Scrap it and start again.

So far, so good. Now what about the Z height reference for G55 workspace, for the second operation, in the ER40 collet? Although I already mentioned the final length of the part was not super-critical, I still wanted to be as accurate as possible, if only for my learning process. I needed to hold the machined end of the part (from the first operation) in a collet. I decided to design the part so its finished diameter from the first operation matched a standard size of collet, thus allowing the collet to hold my piece tightly and accurately without much 'squeezing in' at all. A suitable size for the part in question was 23mm. A collet of that size was fitted into the chuck.

I also added a design feature to my part which gave it a small shoulder, beyond which the diameter was fractionally greater than 23mm, so the part could slip into the collet comfortably but stop at the shoulder and always repeatably in the same vertical position. I also ensured this shoulder was aesthetically pleasing to my design. I figured sometimes you might need to add attributes to a design simply to aid the manufacturing process, rather than for anything the part needed, but it still had to look nice.

In the Work Offset page of PathPilot®, and in G55 workspace, I zeroed the Z axis at the top of the collet, when one of my 'half-made' parts was tightened in it, using the Passive Probe. In my CAM setup, I set the Z point of the origin to be in line with the shoulder feature I had added to the part's design. I figured this would be pretty accurate. I figured wrong.

**The finished part still in the collet chuck after both operations complete**

With the way I had positioned the PART within the STOCK, in the CAM setup of the second operation, I still needed to face off a further 1mm to produce a part which was the requisite 32mm long. And I was finding my parts varying plus 0.15mm and minus 0.00mm, which is way too much error. It seemed clear the collet was not holding my part at the same height each time I placed a new one in. I was sure the shoulder was resting on the collet reliably, so it must be the collet itself was not sitting at the same height each time. And because the finished part was sometimes longer than it should be, the part must have been lower in the collet than I thought.

I investigated this carefully. It was obvious the tighter the large collet nut is turned, the more downward force and the further the collet is pushed into the taper. After all, that is how a collet works, in order to squeeze the part more firmly. But my aluminium alloy part is not particularly compressible and I could not account for the wild variations in vertical placement I was experiencing. And anyway, I was tightening the large nut, although by hand, very similarly each time. It was something else causing this error, and I needed to find the reason.

After much thought, I found the answer lied in the way the collet was being squeezed. The 23mm collet has about 46mm of length and my part was only being held by the first 15mm. This meant the bottom 30mm or so was being forced into the taper with nothing to stop it 'collapsing' and the collet was becoming mis-shaped and out of parallel with the result it was moving further downwards than expected. Minor variations in torque applied to the collet nut were producing marked differences in height, enough to account for the 0.15mm I was experiencing.

My initial 'fix' was to machine a piece of rod, 23mm in diameter, long enough to sit on the parallels underneath the collet chuck and protrude upwards into the collet by about 15mm. In this way, the collet would be held parallel gripping onto my part at the top, and the new rod at the bottom, with a nominal 15mm gap in its centre portion. But when I tried this, I found the tightening of the large collet nut with the spanner felt very strange indeed, almost 'locking' instantly rather than tightening up progressively. Thinking this through further, I realised the action of the collet being tightened MUST allow for some downward movement into the taper, but as soon as it had taken a good grip of the new piece of rod, it could not move down any further as the rod was already standing on the parallels. In fact, the more I tried to tighten, the more I was applying bending pressure on my expensive parallels by the action of the rod being forced downward. A completely undesirable thing to do.

I was close to a solution by this point. All I needed was the ability to have a piece of 23mm rod in the lower end of the collet which was free to float with any movement that needed to be 'taken up' by the action of the collet sliding down the taper as it was tightened.

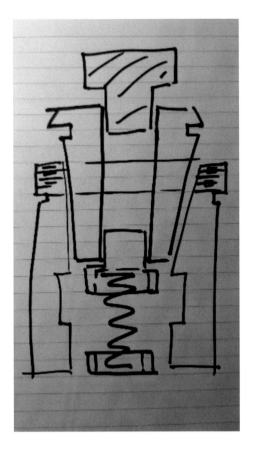

**My rough sketch of the solution for keeping the collet parallel**

I drew a simple sketch of the final design; the 'floating' piece of rod would be held up into the bottom of the collet by light spring pressure, it would have a shoulder so it could not 'intrude too far' into the collet, and I would ensure its diameter would be a 'smidgen' less than 23mm. I chose it to be 22.95mm so this meant the collet would naturally want to grip my part first and this rod second. To all intents and purposes, the collet would still be kept parallel. I don't know if this fine-tuning by 0.05mm really mattered - probably not - but at least I was thinking it through and trying to cover all bases.

I made two parts on a lathe, one the shouldered rod, and the other a sort of thick washer to keep the spring centrally located at the bottom end. I have included a photograph of the separate parts lying on their sides.

Immediately I tried tightening the large collet nut with a spanner, I could tell I had solved this problem – it just felt 'right'. The collet was gripping the part progressively and tightly with smooth spanner action. I placed a PART into the collet and re-probed for the Z axis in G55 by touching off on the top of the collet. This time, even with mildly varying torque on the spanner, the variation in my finished part length was within about 0.02mm which was much better.

I have simplified one aspect in my description, but I will own up to it now. In truth, I could not 'touch off' the top of the collet with my part in it, because the top of my part was much wider than the collet and was shielding it from the probe. So, in fact, what I did was place a piece of 23.0mm bar stock into the collet instead, allowing me the room to manoeuvre the probe into position. What was important, of course, was the collet was tightened onto something the same diameter, as that would ensure the collet was pushed down into the collet by the same amount.

**The parts of my design for keeping the collet in parallel**

And one final detail, the top of the collet is not a sharp edge, the maker has ground a small chamfer on the inside edge so this meant my part, with its very subtle shoulder, rested slightly into the chamfer by a 'few thou'. But it did this by the same amount repeatably, so once I knew the effect, I made a manual alteration to the Z value in the G55 work offset table in PathPilot®, to account for this, which then brought my 'manufactured part length' back to 32.0mm.

I have included these final nuances of detail because I wanted to make the point that designing the 'work holding', or fixturing, is every bit as important as the design of the part itself. Sometimes you might need to add features to the part which are there purely to aid its manufacture. And sometimes, you may find you design a part which cannot be manufactured until you account for and include in its design some aspect which lets you hold it in the machine in a suitable way.

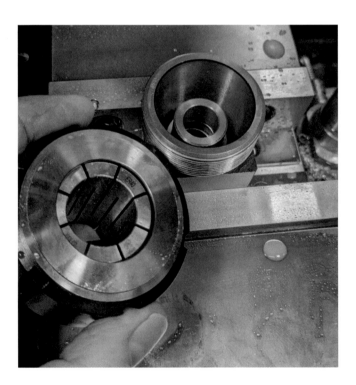

**Close-up showing the sprung 'tube' which seats into the collet**

The sprung 'rod' became a sprung 'tube' when I bored out the centre, so that coolant would escape freely through the whole assembly. This was surprisingly important, as an earlier prototype which did not have this feature proved.

## 13.6 Clamping stock directly to the table

Tormach® produce a 'clamping set' which fits the 5/8-inch slots in the table. It is very similar to the set of tee nuts, studs, washers, hex nuts, clamp bars and support blocks I already use with my small manual mill, only larger in every sense. Clearly there must be a huge factory in China whose sole purpose is making these sets.

I haven't used it yet on the Tormach® 770MX. But I know the time will come when I shall need to. Only, when that time comes, I know I must take care not to crash my machine into the clamps. And that is where you will need to account for the positioning of all clamping in the CAM part of Fusion 360, so the machine knows how high it needs to retract the cutting tool to clear any obstructions.

Sometimes you may require manufacturing features on a flat plate, such as through-holes or slots, which would mean clamping the stock plate on spacers giving some clearance between it and the machine's table. Alternatively, you could clamp the whole plate tightly to the table on a 'sacrificial' plate of material, and this would give it more even support. Since CNC machines are so accurate, this sacrificial plate can be quite thin, even as thin as 1mm, so if you were trying to machine a slot all the way through your part, you could go deeper than the stock bottom by 0.3mm, say, safe in the knowledge you were still 0.7mm away from damaging the mill's table. Obviously, the sacrificial material would effectively be scrap afterwards, so there is a cost implication to this process.

A cheaper option might be to use something like particle board, such as MDF, as the sacrificial layer. Again, you only need to machine slightly into this layer, to avoid any burrs on the finished part. However, this might not be so accurate in terms of repeatability, but for a one-off job, once it is clamped down, and the Z height probed, it would work quite well. It wouldn't be any use, though, if you needed to use coolant, as the MDF would get wet and expand unevenly. And it would be an awful mess.

## 13.7 Use of 'tabs' for holding parts in place

Whether holding large plate material on the table with a sacrificial base, or thicker plate material in a vice, and the plan is to cut out randomly shaped parts, then it may become necessary to 'design in' the use of tabs.

This is a feature of CAM software packages and is calculated to some extent by the program. Essentially it means you can cut out the whole part from a piece of plate material, except for leaving some very small pieces around the perimeter to keep the part from 'falling' out of the stock. To better explain what I mean, you can imagine the injection-moulded parts of a plastic model attached by the thinnest pieces possible to the moulded sprue. In a similar way, the finished part is held in place by just a few 'tabs.' Choosing how many and what size tabs, and where they should be positioned, is a design matter between you and the CAM software. And probably an element of trial and error. Too little support will obviously result in tolerance issues if the part is able to vibrate or chatter.

## 13.8 Advanced fixturing techniques

It gets quite expensive when you start to look at some of the more advanced fixturing systems available. But in return, you are buying very high accuracy, high repeatability, and clever gripping methods for holding stock. One of the downsides of the common machine vice is the natural tendency for the moving jaw to 'ride' upwards when tightened. At 'hobbyist' level using the Tormach® 5-inch vice on a mill table, I do not think this effect is a concern. However, I must point out there are better systems for work holding if you need them. Please look up 'Saunders Machine Works', which is a website and manufacturing company owned by John Saunders, famous for the NYC CNC website and YouTube 'learning centre'. His firm makes 'fixture plates' for many different milling machines, including those of Tormach®. These fix to the table of the mill so that instead of three slots in the X direction, the mill table becomes a much busier place with fixing holes all over. Various devices can fix to this to grip the stock material to be machined in all manner of ways.

# Chapter 14     microARC4

The 4th axis in CNC machining is the rotation of your work around the X axis, and this is labelled the 'A' axis. The microARC4 is Tormach®'s newest 4th axis option for its range of CNC mills. It can be bolted to either end of the machine table, thus facing either way on the X axis, which is something you must, of course, be clear about within your CAM software.

In the case of Tormach®'s microARC4 device, this rotation is effected by a stepper motor, not a servo motor. The control circuitry for the stepper motor is not included in the base machine but is part of the option package and must be fitted by the owner into the electrical cabinet. Note the X, Y and Z axes of the Tormach® 770MX machine are all servo motor driven and, as such, run extremely quietly. When you first connect the microARC4 to the electrical cabinet and test its motion clockwise and anticlockwise in the A axis, you will immediately recognise the almost musical harmonics of the stepper motor doing its thing.

14.1   Hardware in the kit

With the microARC4 unit bolted directly to the table of the machine, the horizontal rotational axis is 80mm above the table. It comes with two work holding options, firstly a 100mm 3 jaw self-centring chuck. This is a nicely made chuck which comes complete with the traditional two sets of normal and reverse jaws. Secondly, an ER40 collet chuck is supplied on a backplate which replaces the 3-jaw chuck. As I already recommended the purchase of a fixed ER40 collet chuck, for using in a machine vice, justification for buying a full set of ER40 collets is easy.

There are some bolts and tee nuts in the kit but unfortunately these do not match up to the tee slots of the 770MX table. I believe the device was designed to fit the Tormach® 440 originally and they don't appear to have included tee nut options for the larger machine tables. If you want to 'go posh', there are fixing systems available online from suppliers such as Saunders Machine Works in the USA which solve this very nicely. But these are quite expensive, and I just needed to bolt the unit down on the table and get to grips with using it. Although Tormach® forgot to include a set of four tee nuts to suit the 5/8-inch slots, they did include some special bolts for the microARC4 unit; I say special since two of them are 6 inches long and clamp from the top. All four of them are 5/16-18 UNC threads which I think of as the American equivalent of the metric size M8.

The decision must be made, then: buy some M8 tee nuts and convert to metric or get some 5/16-18 UNC tee nuts, to fit the table slots. My thinking was, and practical results from searching around confirmed, it was going to be inconvenient to source two 6-inch bolts in M8 (150mm would be cutting it fine). And sticking with the UNC thread, I didn't find it at all easy to find the tee nuts I needed, readymade.

My solution was to purchase 'blank' tee nuts, into which I could cut my own threads. Cutwel produce these and they are referred to as SKU: 1730-16. These fit the Tormach® 770MX table slots rather well and can be drilled and tapped readily. In fact, it is a great way to start learning how to tap with the machine's Conversational rigid tapping mode. I looked for UNC machine taps from Cutwel, who reassuringly stocked a suitably high-quality product: 5/16-inch x 18TPI UNC HSS 5% Cobalt Spiral Flute Universal Machine Tap - TC824 Series (YG-1).

When you look closely at the underside of the microARC4 unit, you will see four slots and two holes. Two of the slots are of no use if you are setting up on a Tormach® 770MX because they are spaced 120mm apart and will not align with the slots of the machine's table. The two front slots are spaced 100mm apart, as indeed are the holes, and these are the four means by which you must bolt it down. It is the holes, not the slots, that take the 6-inch-long bolts which clamp rather securely from the top. All four bolts must be fitted with a plain washer - this is important because the whole of the microARC4 body is made from aluminium alloy and would gall easily.

There is an alignment key/bar which can be secured to the underside of the unit, and this is supposed to engage with the centre slot of the mill table thus aligning the A axis to the X axis. It fits very snugly and works well. However, the first time you set this up you'll want to double-check its trueness by placing a known straight piece of bar into the chuck or collet and sweep along it with a dial indicator.

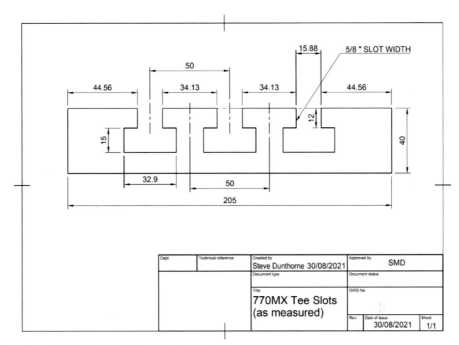

Tee slots on the Tormach® 770MX table

The question will arise: which end of the table do you position the 4th axis unit? The answer is it really doesn't matter in terms of the CAM process but do consider it is quite a tall device and putting it at the left end of the table would preclude effective use of the automatic tool changer, because the tools would simply crash into the box which holds the motor mechanism. The top handle can be removed to help matters if you were only using very short tools, but it would be exceptionally tight for space. You could just empty the carousel completely and resort to manual tool changes.

Alternatively, you could place the microARC4 at the opposite end and use it facing the other way round, which should generate enough space for the automatic tool changer to function normally.

The unit has a protected cable which is long enough to thread through to the rear of the machine and plug into the prepared receptacle on the side of the electrical cabinet. But the electronic control box containing the electronics for the stepper motor driver comes separately in a cardboard box and must be fitted carefully into the main electrical cabinet before the 4th axis will function.

## 14.2 Connecting up the stepper driver control box

When you first open the electrical cabinet with the key and look inside, I believe you will be extremely impressed with the quality of workmanship. I certainly was. The last chapter of the Operator's manual contains the electrical schematic diagrams for the machine including the DIN rail terminal-block labelling system and cross references. It is totally professional, and all matches up clearly and accurately as you would expect.

**Electrical cabinet showing the space (top right) for the stepper driver**

There is a separate instruction set for installing the stepper motor driver board. The wiring which connects to this 'black box' is pre-installed and hidden behind the plastic conduit which surrounds the terminal-blocks, so you must gently prise this open to reveal them. I am not going to discuss the actual procedure – it has already been well documented by the manufacturer. What I would say, though, is even if you are not particularly keen on electrics, this represents no great hurdle and if you follow the instructions carefully, do not be afraid to complete this yourself. Obviously only do this with the power completely removed from the machine. And if you really have 'no clue' about electrics, then ask someone who can help – you want this to work reliably and safely. The wires which are being connected to the box are power supply wires and control signal wires, and any poorly made connections could conceivably upset the rest of the machine. As always, common sense, and indeed safety, must prevail.

Once the stepper control box is wired in, connect the microARC4 unit with its armoured cable to the plug on the outside of the cabinet and power up the machine. Switch on the PathPilot® computer. On the touchscreen there is a secondary tab on the Settings page where you must let the controller know you have fitted the 4th axis.

14.3  Keyboard control

There are two ways to manually 'jog' the moving parts of the machine and both are valid and important. Sometimes, you will need to 'home in' carefully and have your head right inside the open doors. This will best suit the jog-wheel hand control, which is on a flying lead and hangs magnetically from the PathPilot® console, because it allows you fine control without having to reach the touchscreen or keyboard.

However, when you are just 'whizzing' the parts around, moving the table in X and Y, driving the spindle (Z axis) up and down, or rotating the A axis clockwise or anti-clockwise, it may be quite convenient to use the keyboard shortcuts instead. I find them extremely useful and if you choose to use them, you will find that you soon commit them to memory.

Here are the relevant keys for any Tormach® milling machine:

- X axis          LEFT and RIGHT arrow keys

- Y axis        UP and DOWN arrow keys
- Z axis        PgUp and PgDn keys
- A axis        ',<' and '.>' keys (you don't need SHIFT)

All the shortcuts make 'visual' cognitive sense and are therefore easy to learn.


14.4   Alignment of the A axis and centrality of the chuck

The microARC4 unit has no tilt adjustment, so it should be horizontally parallel with the table by default. The A axis must be exactly parallel with the X axis since in the CAM software, the A axis is, by definition, the rotation of the PART about the X axis. In simple terms, the unit must be 'square' to the table. The alignment bar fitted to the underside should ensure everything is in alignment.

Now we must consider if the centre of rotation of the A axis motor is exactly concentric with the centre of the chuck, be that the 3-jaw chuck or the ER40 collet chuck on its back plate. This is normally going to be very important.

It is worth mentioning the fixing of round or hexagonal STOCK in the jaws of a self-centring scroll chuck can never be accurate to the extent we need in CNC milling. That would not be a problem, though, if we were milling our first operation and the rough STOCK was in the chuck. Much the same as when a piece of STOCK is in the chuck of any lathe, so the PART that gets turned will end up being concentric to the centre line of the axis of the lathe even if the jaws are not exactly centred. If we were to consider a second operation, say, when we flip the PART round and wish to machine the other end of it, then is the time we might prefer to use a collet for greater accuracy. In this instance, we would still need to initially check the centrality of rotation of the collet chuck. The set-up manual provided by Tormach® (and available online) discusses how to adjust for any run-out and dial it in accurately.

## 14.5 Fusion 360 and 4th axis work

It used to be the case you could have a Fusion 360 account and the CAM workspace would do all this CNC type work for free if you were a hobbyist making hardly any income from your projects. Then Autodesk changed the pricing structure and quite a few of the 'clever bits' of Fusion 360 were removed from the free version. Two of those were (1) use of an automatic tool changer and being able to generate the code to make the necessary tool change commands and (2) producing any code which utilised the fourth 'A' axis. Hence, it is well worth paying the monthly fee if you are serious about learning these skills.

The pricing tariff gets more complicated and expensive when you get into detailed industrial use including greater number of axes, 5 and up. This won't be bothering us, as the Tormach® cannot run any more than four axes.

This book is not designed to teach 4th axis machining. It is something which I am currently learning myself. It is extraordinarily interesting, and Fusion 360 is well set up for it. There are plenty of resources online to help you. But I wanted to make one important point, just in case you had not come across this idea before. There is a difference between 4th axis simultaneous machining, and something called 3 + 1 axis machining.

The difference is easily explained and obvious. Imagine using a dividing head or an indexing head on a manual milling machine. If you started off with some round bar in the chuck and machined off a flat with an end mill, then rotated the chuck by 60 degrees and milled off another flat and kept doing this six times in all, we can all visualise we would have created something akin to a hexagonal shape.

If we designed the same hexagonally shaped piece in Fusion 360, then ran the CAM software to program the CNC machine to do this, it is reasonably obvious the 4th axis would only be revolving when the milling cutter was NOT cutting. And vice versa. In this case, we would be using 3 axes, when cutting, and using 1 axis, when rotating the part. This so-called 3 + 1 axis machining is simple for Fusion 360 CAM to process.

It takes a lot more mathematics for the same software to calculate the G-code commands for making parts which need the cutter to be cutting at the same time the part is rotating, as I am positive you can imagine. Parts, for example, where you might have a slot around the circumference, that sort of thing.

Luckily, and excitingly, Fusion 360 CAM can do all this. And so can the Tormach®. Full simultaneous 4th axis machining.

**Tormach's 4th axis unit known as microARC4**
Copyright© image courtesy of Tormach® Inc.

# Chapter 15   Modifications and Additions

The following are a few changes I have made to my mill setup. I hope you find them useful. You can ignore this chapter if you wish, and your CNC milling experience will be unaffected. But you know how it is; sometimes you just need to personalise things a bit.

## 15.1   Passive Probe holster

I have written at length about the Tormach® Passive Probe, an option I highly recommend, which enables fast accurate measurement of work offsets in all the axes. When you first connect it to the electrical cabinet using its 'DIN' plug, before it can be used, the manual takes you through a set-up procedure which ensures the vertical centre line through the probing pin is exactly in line with the axis of the spindle itself. It is just a matter of taking some measurements with the probe in various rotational positions and adjusting with a tiny Allen key to centre it in a purely mechanical sense. Once set, in theory it should not go out of alignment unless there is some sort of mishap.

Obviously, its repeatability relies on the BT30 taper of its holder being reliable and true, which it is. The problem comes when you have used the probe, found the work offset data you require, where do you put the temporarily redundant probe? I found myself lying it gently down on its side on the small metal table between the keyboard and the mouse. A BT30 holder and probe is a heavy combination, the whole assembly lies precariously on its side and is vulnerable to getting knocked around. You would be horrified if it were to drop on the floor – it would almost certainly be scrap. It always remains plugged in, and the long cable threads through between the PathPilot® console and the enclosure, which makes it even easier to 'snag' the trailing lead and cause a disaster.

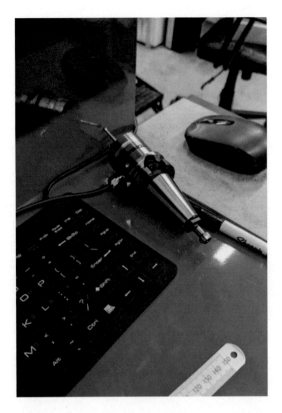

**My Passive probe perched vulnerably when not in use**

It is all, shall we say, a little untidy. I suspect Tormach® have not had the opportunity to address this yet, so I wanted to figure out a quick fix for my own situation. And I designed a holster which clamps onto the large steel pole, which is the main frame of the PathPilot® Console, in much the same way the screen itself clamps to it.

I drew it up in the Design space of Fusion 360. It has two halves which clamp around the pole using two large bolts and nuts. And it utilises a small injection-moulded part readily available on the internet which exactly fits a BT30 taper, the tool holder resting in it upside down. This was probably the lazy option but sometimes that's OK.

I tried printing it myself, but it is quite large, and I wasn't sure of the strength in the Z direction of my print, especially as I was applying tension with the clamping action, so I sent my STL file to an online printing service and had the two parts SLS laser printed in nylon. SLS printing has pretty much uniform tensile strength in all axes, and nylon is very strong.

**Showing my Passive Probe holstered safely**

When you use the probe, there is sufficient cable length to comfortably place the probe into the spindle through the open doors. When you have finished using it, the holster is in an ideal position to place the tool back safely. It works well and I will be happy to share the STL files with anyone who would like to use the idea. The red injection-moulded part 'spring' locks into place with a quarter turn. It is visible in the photograph - these are readily available online for just a few pounds each.

I just want to add one more piece of information about this probe, which is not written down anywhere and is worth knowing. I found out by accident, the probe tip is always electrically active, and the control system continues to monitor it even when it is sitting harmlessly in its holster minding its own business, while the machine is busy making parts behind the closed doors. What happens when you touch the probe during a cycle run? Simple, the machine stops instantly in its tracks and flags up an error on the Status page (the tab goes yellow on the screen) telling you there was an unexpected probe activation. The trouble is, as far as I can find out, there is no way to continue making the part. You must start again from scratch, which is annoying if you knocked it by mistake at the end of a long run. Ask me how I know this?

Tormach® have told me they are aware of this issue, but it seems to be deep-rooted within the original Linux code and, as far as I know, there are no plans to address this anytime soon as it is non-trivial. So long as you are aware, it is not a great problem, and by keeping the probe in a known 'safe place', such as the holster design, this is probably the best way to avoid any further unplanned stops.

Ironically, I only found out about this issue recently, when the probe WAS in the holster, and my sleeve caught the probe tip by sheer bad luck. What can I say?

## 15.2 Handheld safe air-blast gun

When you see operators using large industrial CNC machining centres in factory environments, they always seem to have access to an 'air gun' by their side, for blowing away coolant and swarf from the parts and the fixtures. If you want one with the Tormach® 770MX you will need to set it up yourself. It's completely easy of course; you could just connect any air gun to your compressor and leave it lying on the floor next to the machine, but I put some thought into the whole idea and I shall share this now.

**Safety air gun 3D-printed holding peg**

It turns out there is a square opening and four threaded holes on the front of the left chip pan, just beneath the left door. And Tormach® have added a small metal cover plate with four screws to hide it. I have no idea why it is there, clearly a plan for another day. But it is very useful because it gives an anchorage point for a small plate and peg to hold an air gun on the side of the machine, just about where you'd want one, and fix it with screws which are already there. No need to cut any holes in the enclosure, so no damage done. You can see my air gun hanging on the peg in the photo.

If you are struggling to find a suitable flexible air hose for this purpose, I can recommend one I found online with RS Components Ltd. They stock it as RS Stock No. 700-5658.

I designed and printed a small peg with four mounting holes which accepts the pistol grip of the air gun I had. This immediately struck me as potentially dangerous, so I did a little 'risk assessment' for my own peace of mind. My thought was if a person was to inadvertently brush past this with their thigh or hip, the peg would break off safely, as it was made of printed plastic. If the peg was a steel one welded to a backing plate secured with the screws, it could cause quite a nasty injury. My design is printed in polycarbonate and is strong enough to receive the air gun in a 'boisterous' manner without breaking off, but it certainly wouldn't be able to gouge into human flesh. I am calling that 'fail-safe'.

Continuing in the spirit of risk assessment, on the subject of 'shop air', most of us with engineering backgrounds will appreciate the inherent danger of a 'workshop' air gun. In the domestic situation there may well be children who are free to come and go in the family garage and the idea of a 'live' air supply just sitting on a peg is an obvious concern. I am running my compressor at between 90 and 120psi and I have set the pressure to 90psi at my Tormach® machine using the Filter Regulator Lubricator I mentioned earlier. But if I plug an air gun into my compressor manifold, I could well have 120psi at the gun. In the wrong hands this is, without question, lethal. There is a solution to improve matters and I implore everyone designing their pneumatic supply to consider it.

PCL is a long-standing company, based in Sheffield, South Yorkshire, which produces high quality industrial pneumatic equipment, and they sell an air gun which has a built-in safety device. No matter the air pressure connected to it, the air gun can only ever offer 30psi air pressure at its outlet. Their URL is https://www.pclairtechnology.com/

The health and safety regulations on both sides of the Atlantic now require the use of 30psi pressure restrictors on handheld air guns in industrial settings. But we are discussing hobbyists, garage owners, and 'one-person' start-ups where these sorts of safety advisories might get overlooked or ignored. I believe as soon as you have a pneumatic supply up and running in your own garage, it is time to take things seriously and account for the safety of everyone who may or may not enter that space. Enough said, I trust.

## 15.3 Hours meter

Let me tell you I do love to stick labels on the back of things when they are new, using my self-adhesive label maker. For instance, on the back of my Sage Barista Express coffee machine (the best machine in the house it must be said) there is a tiny white label which simply states

'March 2019', which I put there the day I bought it. Time flies and it is easy to forget how old things are. And I like to know, if a machine goes wrong, is it four or five years old (well fair enough then) or is it only just over a year before failure (disgraceful)? You get the point I am trying to make.

With the CNC machine in the garage, I don't yet know how much I shall use it. I will always know how old it is – I already put a label on it – but I have no way of knowing how hard it is going to work, nor recording the amount of work it has done. I thought, for just a few pounds, it would be nice to fit an 'hours' meter into the electrical cabinet. There is plenty of space remaining next to the newly installed stepper motor control box.

I didn't need anything too technical, such as making sure it only records when the machine is cutting, just an idea of whenever it is powered up would suffice. There were two immediate choices, whether to pick up a mains voltage from the main switch or a low voltage from somewhere downstream of that. With the help of the schematic diagrams and a look around the cabinet, it quickly became clear to me picking up a mains voltage was not really very easy as all the cables are well shrouded and hidden. I found the main 24V dc system within the machine to be much more accessible, even having spare terminals with +24V (Tormach® terminal reference 401) already available for me to connect to. And a 0V return (Tormach® terminal reference 400) was easy to find on one of the contactors. I found a DIN rail mounted hours meter which works on basically anything up to 80Vdc from RS Components Ltd., bought a small piece of DIN rail, and installed it.

Now I fully realise I could switch on the Tormach® mill and leave it powered up for hours on end and yet not actually have it manufacture anything, so I know the hours meter will only give me an approximation of how hard I am making it work. But it is better than wild guesswork and something upon which to base an educated estimate in the future. It is certainly a case of 'fit and forget' to some extent; it might become a useful measure by which I plan forthcoming maintenance, or it might simply prove to be interesting knowledge in a few years' time, but nothing more. On the other hand, it's conceivable it could provide valuable future information if I ever decided to sell the machine on. Time will tell. It's good to keep track of your machines. I am delighted to report my coffee machine is still working wonderfully.

Electrical cabinet with 'A' axis stepper driver and hours meter installed

## 15.4 USB I/O M-code interface kit

On the pricelist, the USB I/O option was inexpensive, and I appended it to my purchase order, not really knowing anything about it, or what I would do with it. But for the minimal extra cost, it seemed worth a punt. It has turned out to be rather excellent and I am delighted with it. For months it was lying in its cardboard box on my bench and one day I took a moment to connect it up and see what it could do.

There are four inputs which it can detect electrically. I don't use these. It is a bit advanced for me to think up anything I would ever need to do with my Tormach® which would involve 'testing' the condition of an input. Never say never I suppose. But so far, I personally haven't come up with any use for these inputs.

There are also four outputs. This is the bit which is more useful for someone at my level. Each of the four outputs are presented to the user as simply as it could ever get: namely, the three terminals of a single pole, double throw 'changeover' relay contact set. They are totally isolated from anything else in the box, and you are quite safe to put mains voltage on them if you wish, without fear of damaging your PathPilot® computer. The logic in the box itself is powered through the USB lead that connects it to the computer. It could not be simpler.

The four outputs are screw terminals, Normally Open (NO), Common (C), and Normally Closed (NC), for each one.

Although I indicated it would be OK to join mains voltage to the box, it wouldn't be considered the normal way to go about things. In machines, the best practice is to run control gear at 24V dc which is much safer. And you can purchase small DIN rail 'switched mode' power supplies to produce it. These are great because they 'auto-shutdown' if there are any short circuits or nasty faults. As the USB I/O box is itself designed to fit on a DIN rail, I decided to get some more rail, hook up a few terminals on it, and make things a tad more professional than just having the box lying on the floor with some wires hanging out of it.

I needed to quickly decide how many of these outputs I was going to need.

## 15.5 Signal tower

In my pursuit of the industrial look, having already bolted an air gun to the side of my Tormach®, I clearly needed to get hold of a 'signal tower'. You may initially laugh out loud, but it has proven itself to be extremely useful. For those wondering what I mean, the signal towers are lamp fittings you often see attached to the top of machines which can show many colours such as red, green, amber etc to indicate across a factory floor the condition state of a particular machine. For example, 'process complete', a 'fault condition', or 'operator attendance' required or something like that. A similar sort of thing is often used above each checkout at supermarkets.

**Signal Tower attached to the top of the Tormach®**

I bought one from RS Components Ltd., which only had two coloured lamps because I couldn't think of that many things I wished to be warned about. They have LEDs inside, but being industrial fittings, they are powered by 24V dc which was ideal. My new RED and GREEN signals would now account for two of my output channels on my USB device.

Once the PathPilot® controller knows about the presence of the USB I/O device (you must tick a box on the Settings page) you can test the outputs by using the touchscreen buttons on the Status page. You can hear the relays clicking in and out. It is basic stuff. Where it gets more useful, though, is through software control of each output and, for this, we must learn some more G-code. In fact, to be pedantic, we will be using M-codes, which I already mentioned before but here's a recap.

Remember the M codes are 'actions', such as M6 (change a tool) or M8 (turn on the coolant). With the four outputs on the USB I/O device, we can now instruct the control software to turn on, or turn off, each of them and the code is a simple M64 for on, and M65 for off. The four outputs are labelled outputs P0, P1, P2, and P3.

Here's an example of a line of code:

    M64 P1    (turn on the second output channel relay)

I decided to connect my RED lamp in the signal tower to the normally open (NO) terminal of the first output, called P0, and my GREEN lamp to the normally open (NO) terminal of the second, P1. And I provided 24V dc to the common terminals of both. In this way, both lamps would normally be OFF, like at initial power up and such-like.

**Temporary 'hook-up' of the USB I/O device on my DIN terminal rail**

I don't need to use these all the time, of course. But, at one stage, I was making quite a few parts which took about 40 minutes fully automatic operation for each one, after which I needed to attend the machine and insert fresh STOCK for the next part. I decided, once I had finalised the G-code for the operation, to insert some of my own code at the beginning to put the RED lamp on, and then at the end of the code, switch over to just GREEN on. In this way, I could leave the machine going, and every time I passed nearby, I could see, with a casual glance, immediately whether the operation was finished or still going. It works superbly well.

## 15.6 Pneumatic solenoid for remote air blast

I thought up a use for the third output on the USB I/O system; remote fixed-position air blast for the table. This is for those occasions you don't need, nor desire, flood coolant. For instance, maybe you are doing some ad-hoc work on a piece of MDF wood, or some plastic, and an air supply might be better suited to the job.

The normal G-code for Flood Coolant ON is M8. There is another code, M7, which turns on a separate coolant channel and this can be used for coolant mist systems which Tormach® can supply. A quick glance in the back of the electrical cabinet, and it is clear this channel exists on the main control board with separate relay control. I might still purchase the 'mist' option in the future, so I decided to leave the 'M7' channel unused for now and use the USB I/O channel 'three' for my separate 'dry' air system instead.

It is important to know that both M7 and M8 coolant controls are cancelled by an M9 command. These are accounted for in a drop-down list within Fusion 360 post processor, so you must ask Fusion 360 to make your choice of coolant, switch it on and off at the appropriate time, within the post processing setup. Adding commands for the output function of the USB I/O board from within Fusion 360 - as far as I know - is not possible. Advanced 'programmer' types may know better but I am sticking with what I know for now and will simply add some code in the edit afterwards. All I will need are these two commands in suitable places, so the air supply is switched on when cutting, and off again when no longer needed:

    M64 P2    (air blast solenoid ON)

    M65 P2    (air blast solenoid OFF)

For 'ad hoc' prototyping work, it is also possible to manually switch the air blast on using the touchscreen button on the Status page of PathPilot®, but only before or after a milling operation. The controls on this page are not active whilst running a cycle. Because of this, I have also added a manual override switch, a microswitch in a box, attached to the front of the machine, which bypasses the third channel relay contacts and opens the air valve with a 'shot' of 24 volts.

For the mechanicals, I have run a flexible pneumatic line to a nozzle which has a tap to adjust the air flow. This is part of an assembly fixed to the table with a bolt and tee nut, suitably positioned to avoid the possibility of collision with the spindle or the tool changer. The other end of the line protrudes through the back of the machine to a 24V dc solenoid-operated pneumatic valve (RS Stock No. 892-9993) which I have mounted on a fabricated metal bracket at the rear of the Tormach®. This is fed by a separate Filter Regulator (not a Lubricator) which in turn is supplied by my main compressor manifold. In this way, I have separate control of the pressure to this line and some control of the flow rate through the nozzle using the tap. I connected the solenoid to channel P2 (which is the third channel).

# Chapter 16   Learn from Others

I made a big thing about my need for research, both from reading books and scouring the internet, to put me in a position ready and suitably qualified to purchase the Tormach® 770MX mill. That process never stops, rather it gets more and more compelling and, as your knowledge grows, your thirst for it does too. And the funny thing is, the more skill you acquire, the better you get at knowing where to find the information you need. It is a self-fulfilling cycle.

I wanted to tell the story of my journey of how I got this mill up and running and making great parts because I believe there are many others out there who are considering doing this but are not sure if they can. There is no doubt about it, things these days aren't simple. Stuff is complicated. It's just the way things are. You must 'put in' a modicum of work to gain the necessary knowledgebase to effectively manage something like this project.

But the information is out there. And it's magnificent in its vastness. When I was growing up, adults would go to 'evening classes' or 'night school' to learn new hobbies and skills. Nowadays, of course, we have YouTube which I regard as one of the most important 'educators' in the modern world.

Of course, I hope there is still room for the written word. I certainly believe there is, and so I begin this last chapter, which is a roll call of all the people who have written a book I found useful; offered some advice on a forum I chose to take note of; taught me on a 'one-to-one' basis in their carefully scripted YouTube tutorials; chatted about their own pet projects online; or even work in industry and still kindly took the time to answer my emails. Everyone I can think of who has helped my journey, and every company I can think of which has been eager to engage with me, or simply sell me good products, I shall try to list here. In a sense this is a combined 'online' bibliography and thank you.

## 16.1   Lars Christensen

Lars Christensen is an experienced designer and CNC milling expert, well established as an online educator with an extensive range of YouTube tutorials available for anyone to watch. I believe he is employed by Autodesk and is an expert tutor in the use of Fusion 360 CAD/CAM software. He comes first on my list because when I bought a desktop 3D printer, his videos were the first I watched and could understand. If you are truly new to 3D design and modelling, as I was, search for "Lars Christensen Absolute Beginners Part 1" and watch his video about drawing a small plastic 'conduit' box and lid. He comes across totally at ease and chilled but, crucially, he teaches at a sensible speed and gives you that first 'in' that you need. It draws you in – no pun intended - and soon you realise you can draw in three dimensions. He clearly loves his subject and is a brilliant instructor. If I have a problem which I am finding hard to solve in the 'parametric modelling space' or the 'CAM manufacturing space' of Fusion 360 I will 'tack' his name on the end of my Google search string, hoping he has done a video on the issue, and often he has. Don't worry that some of his early videos were made before Autodesk upgraded the graphical user interface a little bit; it doesn't matter that much, and you will still learn how it works. The man is a legend in my view.

## 16.2 John Saunders

Another legend in this field is John Saunders who set up from scratch in New York with a small mill in his basement and started the extremely successful YouTube tutorial site called 'NYC CNC', then later moved away from the city, and expanded into 'shop work', including making fixture plates for milling machines under the brand name 'Saunders Machine Works.' Like Lars, John is clearly an extremely clever man, who just loves everything to do with machining and has the same knack of imparting his knowledge and making it fun. His name is another one for adding to the end of your Google searches when you are trying to figure something complicated out. He has made many videos about Fusion 360, and 'feeds and speeds', and his website is an Aladdin's cave of useful data. Go there, watch his videos, you will be there for hours, trust me. John is one of the ambassadors for Tormach®® and I believe still runs a Tormach® mill, although he also now runs larger machinery too. I can honestly tell you that at least seventy-five percent of all I have learned in the last nine months about CNC milling and Fusion 360 came from Lars and John.

## 16.3 Mark Terryberry

There is another great 'character' residing within the YouTube world of engineering and his name is Mark Terryberry who produces very professional (and amusing) training videos for the use of Haas milling machines, for whom he works. I have learned such a lot from Mark, too, so maybe I had better make that earlier seventy-five percent estimate a bit less. Another person born to teach; he has been given a decent budget to produce high quality material. Some of his content relates specifically to Haas equipment but that doesn't matter at all, there is still much to learn from him. He is tenacious, despite being USA based, at catering for both the imperial 'inches' brigade and those of us on the other side of the Atlantic counting in millimetres. In fact, when the subject is crucial and complex, I have seen cases when he has made separate videos on the same subject, one for each of the two measurement systems. How generous is that!

## 16.4 Cliff Hall

Look up Cliff Hall online. He produces his own YouTube videos and makes parts under the brand name Hallmark Design Ltd in New Zealand, using various workshop machines including an early version of the Tormach® 1100 mill. Clearly a talented engineer, his videos are full of useful advice and knowledge. He seems to have a separate YouTube channel named 'Threadexpress' so look that up too. There is much more detailed engineering content in Cliff's videos than you first realise and you may well be rewarded by watching each of his videos more than once, as I was.

## 16.5 Joe Pieczynski

Another YouTube 'must-watch' character is Joe Pieczynski. His glorious accent sounds to me like he is from either New York City or New Jersey or something like that, but I believe he operates his engineering business out of Austin, in Texas. Although he is skilled in all workshop disciplines, including CNC milling, what I find most interesting are his videos on manual lathe and mill skills, and just breaking down simple machining principles into the basic building blocks, explaining what he is about to do, he then goes and does it. He produces some of the best engineering videos I have seen. And you learn something every time you listen to him.

## 16.6 Tormach® online

While I am listing my favourite YouTube channels, and in the process recommending them to anyone thinking of taking this journey into the CNC world, I must not forget to mention Tormach® themselves, since they have produced some excellent video tutorials which are obviously about their own products, but also excellent in a generic sense at teaching some of the basics of CNC milling.

## 16.7 Fusion 360

Autodesk Fusion 360 is a YouTube channel operating in parallel with the Fusion 360 software maker. Oftentimes my Google searches have ended up with my watching one of their video tutorials, which are very well made. They can be a bit quick sometimes, so you find yourself clicking on the screen to activate 'pause' now and then, while you 'go back' to your software to try what they are suggesting. But this is fantastic teaching, at the very highest level. It is backed up by an excellent 'forum' too, where questions which come into your own head have already been asked by someone previously. The answer is always in there somewhere.

## 16.8 SkyCAD Electrical

SkyCAD Electrical CAD drawing software is free to download and use for personal use. Based in Montreal, Canada, I found their system very good at producing simplified electrical drawings, suitable for inclusion in this book. The software was very intuitive and clearly aimed at producing far more complex diagrams than I needed on this occasion. I wish to record my thanks to them for the use of this software.

## 16.9   Jayson Van Camp

There is a gentleman called Jay at www.vcedgeknives.com and he is a talented engineer with years of aerospace engineering experience. He now makes knives, rather beautifully crafted knives it must be added, with the help of his Tormach® 1100MX milling machine, which is 'one up' from my own 770MX. He has a very nice 'talking to camera' manner and has explained his use of the Tormach® very well indeed, so for that I recommend a visit to his website and YouTube channel.

## 16.10   Titan Gilroy

Another name you will come across is Titan Gilroy who owns 'Titans of CNC' YouTube channel and machine shop. A troubled past and some time in a state penitentiary as a guest of the United States Government, Titan tells the story of his ultimate escape from gangland fighting by 'finding' CNC machining. A natural talent and orator, he is a passionate advocate for the American youth being given the opportunity to be taught how CNC works just like he was. He runs an academy which is free to join online and some of his videos are absolutely packed with information useful to us as newcomers to the subject. If you find his speech to the American Industrialists at a lunch function, online, about his passion for getting the USA to make things again, rather than buy them from China, I hope you will find it as emotionally charged as I did.

## 16.11   Norman Kowalczyk, Daniel Rogge, and Andrew Grevstad

I want to highlight three particular people within the Tormach® company who have been exceptional in helping me through some teething difficulties as I assembled and ultimately 'got to grips' with using my Tormach® 770MX. I mentioned before in the book Tormach® are very open to listening to your technical problems and trying to help you solve them. I found myself having many email 'conversations' with a tremendous engineer called Norman Kowalczyk, who never seemed to tire of the 'snags' I was experiencing. Together we fixed all the tiny problems which are inevitable in a project like this. I wish to record my heartfelt thanks to Norman for his perseverance and technical skill and knowledge, and just for being a very pleasant person to deal with.

Norman, if you weren't so far away, I would buy you many beers. By now, you are building a picture in your mind of Tormach® being a company which truly values its customer base. After all, as long as I still own my Tormach®, they say they will always try to help me with any problems. That culture comes from the top, and the CEO is currently Daniel Rogge. I had a few email exchanges with Daniel and wish to add my thanks to him too, for his exceptional customer service. You will find Daniel in many of the videos that Tormach® have produced online. And thirdly, Andrew Grevstad, who is Business Development Director, is an excellent ambassador for Tormach® and I continue to keep in touch with him. I believe Andrew was with Tormach® from the very beginning, so knows everything there is to know, and he has certainly been of great help to me in producing this book, so a huge thank you to him too.

16.12   John Grierson

I would like to thank John Grierson, Sales Director at Pennine Lubricants Ltd which is based in Sheffield, South Yorkshire, England. When I was researching coolant oil for the machine, I really did not know anything at all, and I approached Pennine Lubricants for help. I clearly set out from the beginning I would only be buying a few tens of litres, and I wasn't an industrial customer wanting hundreds of gallons, yet John took the time to explain many of the intricacies to me and recommended a particular blend I should benefit from. I am a loyal customer of his now, and I remain grateful for his time and advice. I wholeheartedly recommend this company for coolant oil and 'ways' lubricating oil.

16.13   John Ackerman

I also wish to give sincere thanks to a gentleman called John Ackerman, who is a sales engineer with Blakley Electrics who are based in Crayford, Kent, England. They predominantly manufacture mains electrical transformers for industrial use. I had extensive conversations with John, both on the telephone, and by email, to glean as much information as I could about step-down mains transformers and, ultimately with John's guidance, choose a suitable unit to power my Tormach® machine. John, from the outset, totally appreciated the importance of setting up a robust power supply for operating a 115V machine in the UK, and indeed the rest of Europe, which is resilient to the complications caused by 'residual current' safety devices and offered me professional advice. I was humbled when he chose to set up a specific part number for a transformer which would solve the issue for anyone intending to install a Tormach® 440 or 770 range machine, and it is my sincere hope he will sell many more of these transformers, as a direct result of being so helpful to me when I needed it. Secondly, of course, anyone deciding to install a Tormach® 1100 range machine, which operates on 230V, might do very well making contact with John to discuss options for an isolating transformer with no voltage change to solve those same complications, and he assures me he would be delighted to receive enquiries.

16.14   UK model engineering magazines

The two sister magazines 'Model Engineer' and 'Model Engineers' Workshop' fall under the one umbrella website and their forum resides at this URL address:

https://www.model-engineer.co.uk/forums/

There is quite a lot of useful information to be gleaned from a little searching. Try typing 'Tormach®' into their search box and there is plenty of useful advice freely available. It is friendly too, and I send my thanks to a few people on there for engaging with me on a few occasions. I believe both magazines are UK based. I have certainly been an active reader of them over the years.

16.15   Autodesk Inc.

I would like to recommend a free online publication called 'Fundamentals of CNC Machining' which is published by Autodesk, who are the corporation producing the CAD/CAM software called Fusion 360, which I now know to be an amazing piece of cloud-based software. If you type that title into a Google search box, you are bound to find it. It is downloadable as a pdf file and can be viewed on your desktop, an iPad or some such tablet, or indeed you could get it printed. It is a great starter book to explain the basics of CNC for those who are struggling with some of the three-dimensional concepts. They have made a point of keeping the diagrams as simple and clear as possible. And it's free.

16.16   Joe Evans

I found information in an article written by Joe Evans on a website to do with motorised pumps.

https://www.pumpsandsystems.com/pumps/motor-horsepower-torque-versus-vfd-frequency

I contacted Joe and asked him if I could reproduce his theoretical graph for a power torque curve when a motor is 'over-driven' by a variable frequency drive system, and he kindly agreed. I went on to read some of his presentations on motors on his own website, which I heartily recommend if you find the subject of interest. You can find him at: http://www.pumped101.com/index.html#pump_ed_101

It certainly helped me to appreciate how the spindle motor is being driven, and I thank him for allowing me to share this.

16.17   Rob Joseph

Rob is a colleague I met through work who first awakened my interest in 3D modelling, desktop printing and ultimately CNC machining. He is highly experienced in this field and, indeed, teaches many of the aspects professionally. I thank him sincerely for sowing those initial seeds in my head.

# Chapter 17 Formulae and G code summary

Vc   -   Cutting speed (m/minute)
Fz   -   Feed rate per tooth (mm/tooth)
n    -   Spindle speed (rpm)
π    -   pi (=3.142)
D    -   Diameter of tool (mm)
Vf   -   Feed rate (mm/min)
Z    -   Number of teeth, or flutes
Fn   -   Feed rate per revolution (mm/rev)
MRR  -   Material removal rate (mm^3/s)
Pc   -   Cutting power (W)
Ae   -   width of cut (mm)
Ap   -   depth of cut (mm)
SCE  -   Specific cutting energy (J/mm^3)
T    -   Cutting torque (Nm)
w    -   Angular velocity (Rad/s)

$$n = \frac{Vc \times 1000}{\pi \times D}$$   ; spindle speed

$$Vf = Fz \times Z \times n$$   ; Feed rate (milling)

$$Vf = Fn \times n$$   ; Feed rate (drilling)

$$MRR = \frac{Ae \times Ap \times Vf}{60}$$   ; Material removal rate

$$Pc = \frac{Ae \times Ap \times Vf \times SCE}{60}$$   ; Cutting power

# G code

| | |
|---|---|
| G0 | Rapid positioning |
| G1 | Linear interpolation |
| G2 | Clockwise circular interpolation |
| G3 | Counter-clockwise circular interpolation |
| G4 | Dwell |
| G17 | Plane selection for circular interpolation |
| G20 | Imperial machine mode (inches) |
| G21 | Metric machine mode (millimetres) |
| G28 | Predefined position |
| G28.1 | Store G28 position |
| G30 | Predefined position |
| G40 | Cancel cutter radius compensation |
| G43 | Apply tool length offset |
| G49 | Cancel tool length offset |
| G53 | Move in absolute machine coordinate system |
| G54 | Work Offset 1 (also called 0 in Fusion 360 WCS system) |
| G55 | Work Offset 2 |
| G56 | Work Offset 3 |
| G57 | Work Offset 4 |
| G58 | Work Offset 5 |
| G59 | Work Offset 6 |
| G59.1 | Work Offset 7 |
| G59.2 | Work Offset 8 |
| G59.3 | Work Offset 9 |
| G61 | Exact Path Control |
| G64 | Blended Path Control |
| G73 | Canned cycle – peck drilling with small retract |
| G76 | Multi-pass threading cycle |
| G80 | Cancel motion mode (including canned cycles) |
| G81 | Canned cycle – drilling |

| | |
|---|---|
| G82 | Canned cycle – drilling with dwell |
| G83 | Canned cycle – peck drilling with full retract |
| G85 | Canned cycle – boring, no dwell, feed out |
| G86 | Canned cycle – boring, spindle stop, rapid out |
| G88 | Canned cycle – boring, spindle stop, manual out |
| G89 | Canned cycle – boring, dwell, feed out |
| G90 | Absolute distance mode |
| G94 | Normal feed mode, in millimetres (G21) per minute |

## M code

| | |
|---|---|
| M0 | Program stop (needs cycle start to continue) |
| M1 | Optional program stop (needs cycle start to continue) |
| M2 | Program end |
| M3 | Rotate spindle clockwise |
| M4 | Rotate spindle counterclockwise |
| M5 | Stop spindle rotation |
| M6 | Tool change command |
| M7 | Coolant on (other types of coolant – separate channel) |
| M8 | Coolant on (flood coolant) |
| M9 | All coolant off |
| M10 | Unclamp automatic collet closer |
| M11 | Clamp automatic collet closer |
| M30 | Program end and rewind |

The following used with the USB M-Code I/O Interface Kit option:

| | |
|---|---|
| M64 | Activate output relays |
| M65 | Deactivate output relays |
| M66 | Wait on an input |

# Chapter 18   Suppliers

My goodness, you could write an entire book about cutting tools. But of course, that isn't the intended main thrust of this book. Let's summarise where we are. We have a highly accurate CNC mill, some expensive BT30 spindle tool holders with a useful selection of ER20 collets, and absolutely no cutting tools!

We want to buy decent quality tools because we know the Tormach® is worth it, but on the other hand we fully appreciate we don't need the finest quality tooling you would fit in a half million-pound CNC Machining Centre.

Similarly, we agreed we wouldn't raid the tobacco boxes in the garage containing some old blunt twist drills which have spent half their life in a Black & Decker power drill and nor would we consider those worn-out end mills we last used on the Warco mill facing off some pieces of scrap steel. We need to find the right balance.

Where can we go? Who can we ask? What do we need? As you can imagine, I have done my fair share of research on this and come up with a handful of companies in the UK which I found particularly helpful.

Tormach® Inc.                                              (Many accessories)
4009 Felland Road, Ste. 120
Madison
WI 53718
USA

Phone: +1 608 849-8381
https://Tormach®.com/

I have found ordering from the USA online purchasing site a little laborious as they are not really set up for automatic calculation of shipping to other than USA destinations yet, so you are forced to request a quote, then email them telling them where you are. But all is well, from then on, the process is reliable and quick, using DHL logistics. Some of their accessories are superb. I particularly recommend their BT30 ER20 collet chucks for high quality but small gauge height, which is advantageous.

Protool Ltd                                    (Cutting tools)
Unit 1 Invicta Business Park London Road
Wrotham
Sevenoaks
Kent
TN15 7RJ
Phone: 0844 335 8984
Email: sales@protool-ltd.co.uk
URL: https://www.protool-ltd.co.uk/


Cutwel Ltd                                     (Cutting tools)
Unit A Riverside Drive
Cleckheaton
BD19 4DH
Phone:      0333 006 8532
URL: https://www.cutwel.co.uk/
(recommended contact: Jordan Hyde, Account Manager)


Associated Production Tools Ltd                (Cutting tools)
Unit 7
28 Queen Elizabeth Avenue
Hillington
Glasgow
G52 4NQ
Phone:      0141 892 1010
URL: https://www.shop-apt.co.uk/


Trucut Tools                                   (Cutting tools)
45 Cobham Road
Ferndown Industrial Estate
Wimborne
Dorset
BH21 7QZ
Phone: 01202 717 110
URL: https://trucuttools.co.uk/

Pennine Lubricants                                  (Coolant mix)
Pennine Lubricants Limited
32 Atlas Way
Sheffield
S4 7QQ
https://www.penninelubricants.co.uk/


Blakley Electrics                                   (Transformer)
Southern Service Centre
1-3 Thomas Road
Optima Park
Crayford
Kent
DA1 4QX
https://www.blakley.co.uk/


SGS Engineering UK Ltd                              (Engine hoist)
1 West Side Park
Belmore Way
Raynesway
Derby
DE21 7AZ
https://www.sgs-engineering.com/


Pallet Truck Warehouse Ltd                          (Pallet truck)
78 Alston Drive
Bradwell Abbey
Milton Keynes
MK13 9HG
https://pallettruckwarehouse.co.uk/

© RS Components Ltd.  (Engineering supplies)
Birchington Road
Corby, Northants
NN17 9RS
UK
https://uk.rs-online.com/web/

Most definitely worth having an account open with RS Components for those 'engineering solutions' which you need in a hurry. Quality merchandise, although quite expensive, but you must pay for convenience.


LANTRO JS  (BT30 toolholders)
Amazon UK seller


I AND I International Co. Ltd  (coolant pump upgrade)
Taiwan (R.O.C.)
http://www.iniint.com/


CNC Machine Tools Ltd  (Tormach® importer)
Unit 8s
Chalk Lane
Snetterton
Norfolk
NR16 2JZ

Online URL: https://www.cncmachinetools.co.uk/Tormach®/ and this is their link to the Tormach® importers page. I believe CNC machine Tools Ltd are the sole approved distributor for Tormach® machines in the UK and some other European countries. Mr Eddy Clapham is the managing director. He is very 'hands-on' and helpful. I needed to make a small metal ramp to get the delivered crates into my garage from the drive, and he personally arranged for his company to fabricate this from my drawing, and it arrived with the delivery. You can't say fairer than that for outstanding customer service.

All the above companies have extensive websites listing all manner of cutting tools, a lot of useful data, and high-resolution photographs to help visualise the detail, and I found it educational to spend a good deal of time investigating them in fine detail.

It was evident these suppliers are keen to help with technical advice if you are struggling with some of the finer points. I can put my hands up high and admit I had no idea how complex the world of cutting tools had become. If you email any of them for advice, you will not wait long for help.

## Notes

# Notes

# Notes